THE CURE IN
THE CODE

THE CURE IN THE CODE

HOW 20TH CENTURY CENTURY LAW IS UNDERMINING 21ST CENTURY MEDICINE

PETER W. HUBER

BASIC BOOKS

A Member of the Perseus Books Group
New York

Published by Basic Books,
A Member of the Perseus Books Group

Books published by Basic Books are available at special discounts for bulk purchases
in the United States by corporations, institutions, and other organizations. For more
information, please contact the Special Markets Department at the Perseus Books
Group, 2300 Chestnut Street, Suite 200, Philadelphia, PA 19103, or call (800)
810-4145, ext. 5000, or e-mail special.markets@perseusbooks.com.

Designed by Trish Wilkinson
Set in 11 point Goudy Old Style

Library of Congress Cataloging-in-Publication Data

Huber, Peter W. (Peter William), 1952– author.
    The cure in the code : how 20th century law is undermining 21st century
medicine / Peter W. Huber.
        p. ; cm.
    Includes bibliographical references and index.
    ISBN 978-0-465-05068-0 (hardcover) — ISBN 978-0-465-06981-1 (e-book)
    I. Title.
    [DNLM: 1. Biomedical Research — legislation & jurisprudence—
United States. 2. Drug Industry— legislation & jurisprudence—United
States. 3. Legislation, Medical—United States. 4. State Medicine—legislation &
jurisprudence—United States. W 33 AA1]
    RS380
    338.4'76153—dc23   2013026302

10 9 8 7 6 5 4 3 2 1

For Theodorus and Dorothea and the gifts that followed

CONTENTS

PART FOUR   THE CULTURE OF LIFE

WETWARE

NATURE TOOK FOUR billion years to develop the code that choreographs life on earth today. Medicine's transformation into an information industry began around 1850 and crept forward at a glacial pace for a century. Then in our generation, in the blink of an eye, medicine plunged into the molecular abyss, and in doing so it became all information, or close to it. Unleashing the enormous power and economies of innovation on this, the last frontier of the information revolution, will require fundamental changes in public policy.

Humanity, as represented by the Nobel Committee, officially recognized the dawn of the new era on October 18, 1988, forty years after Gertrude Elion and George Hitchings began their pioneering work on "structure-based" drug design. Elion and Hitchings, two pharmacologists working for Burroughs Wellcome, sought to systematize drug development by studying the molecular structure of the germ or cancer cell and then designing a mirror-image molecule to target it precisely. When they started collaborating, the tools needed to scrutinize life that closely and assemble mirror-image biochemicals were just beginning to emerge. Today, drug designers can read every single molecular letter of life's code. And while designing drugs to control that code isn't easy, the process is, at its best, as inherently logical as the one that composes the software used to control your computer. Such drugs have the power to control almost any disease caused by our own rogue biochemistry or an infectious microbe.

Down at the very bottom, life is launched and propelled by two groups of big, complex molecules—nucleic acids assembled into genes, and the proteins that those genes define. All life uses this bootstrap code to lift itself up from the stillness of inanimate matter. Four molecular letters, arranged in groups of three, define sixty-four words, which are translated into twenty amino acids, along with one punctuation mark to tell other molecules where to start reading and where to stop as they assemble the amino acids into proteins that assemble other molecules to create functional life.

When Elion and Hitchings started developing drugs, almost nothing was known about this code, and science could scarcely imagine the possibility of reading and editing it systematically to control human health. That had changed by the time they arrived in Stockholm. In three fantastically innovative decades, medical science had acquired the tools that would give medicine the power to diagnose and treat from the bottom up. Restriction enzymes: scalpels for genes, infinitely finer and more precise than anything ever handled by an eye surgeon; Nobel Prize, 1978. Recombinant DNA: sutures for genes, sew them back together any way you like; Nobel, 1980. Monoclonal antibodies: Gutenberg's printing press for proteins—set them in biochemical type, run off as many copies as you need; Nobel, 1984. The polymerase chain reaction: Gutenberg's printing press for genes; Nobel, 1993.

The digital crowd uses the term *wetware* to refer to the human brain, but for our purposes wetware is code stored in biochemicals rather than on a hard drive—code so smart that micrograms of it can assemble and control hundreds of pounds of dumb flesh. Like software, wetware can be erased, corrupted, infected, hijacked, and edited, letter by letter, word by word. Parts of our immune system subtly reconfigure their own code on the fly, in random ways; so do cancer cells. Good code keeps us healthy; bad code makes us sick. Medical and crime labs read biological code to diagnose and identify. Biochemists design good code to tweak, tame, patch, or erase the bad.

The power to manipulate biochemical code at will has implications for every sector of our economy. Healthier workers are more productive. Wetware technologies also allow us to bioengineer better crops and reengineer bacteria to extract oil from tar sands or sift gold from sludge. Inevitably, these technologies will also be used to craft dreadful biological weapons—and, if we stay ahead of the curve, defenses against those weapons.

The power in nature's code and our mirror-image drugs resides in molecules that, with technologies now in hand, are as easy to read, copy, and manipulate as silicon chips, or soon will be. Wetware technologies are improving even faster than their digital siblings. Wetware knows how to move and handle and process not just bits but stuff. Toss Intel's best microprocessor into a sandbox and watch what happens: nothing. Intel's silicon doesn't melt more silicon out of the sand and assemble a trillion copies of itself. Wetware does that kind of thing inside us and around us all the time. It grabs material and energy from its surroundings to build copies of its own hardware—hence our green planet, covered with life.

Wetware technology, which came of age about twenty years behind the hardware and software of Andy Grove, Steve Jobs, and Bill Gates, will eclipse everything that the information age has delivered so far. After all, it created Andy, Steve, and Bill. It will also be at least as unruly and disruptive as any of their creations has ever been. Most immediately, the methods, imperatives, and economics of wetware technology can't be squared with entrenched views about how health care ought to work.

———

A PROTEIN THAT choreographs many of the changes that occur in a woman's body in the course of pregnancy resembles a similar protein in a frog. Not so very long ago, medical labs injected frogs with the urine of aspiring human mothers; if the frogs laid eggs, the women knew they were pregnant. The herpes virus that lurks in over 70 percent of humanity answers to the call of another protein produced by human nerve cells near the ear. After creating a first round of nasty sores, the virus hibernates for years, then suddenly wakes up and attacks again when stress, sunburns, or fevers stimulate production of that one protein.

Code is the only suitable word for molecular-scale autocrats that are smart enough to govern pregnancy, viral attacks, and all the surrounding mountains of mindless matter and energy that we call life. We share much of our code with much of the rest of life on earth—parts of which know how to read and hijack our code for their purposes, just as we know how to hijack a frog's for ours.

Disease is hostile code sabotaging friendly code. Cholera, for example, is a short strand of alien code hijacking, among other things, the code that controls the flow of water into our intestines. As we come to understand such processes, the lines we've previously drawn between "health" and "disease" get blurry. Molecular code that helps keep us healthy or just lies dormant today may be enlisted in a process that harms us tomorrow, as cancers and autoimmune disorders routinely do. The retroviruses that cause AIDS and one form of leukemia splice their genetic code into our own DNA. Prions are single molecules—proteins—coded in genes or transmitted like germs; they transform other proteins in humans and cows into more prions in a lethal chain reaction that leaves the host's brain riddled with holes or plaques. Cancer cells, as described by Sherwin Nuland in *How We Die*, are like teenage gangs that roam aimlessly and trash their own neighborhoods. We now understand many of the molecular details that make the cancer cells' chemistry—which is *our* chemistry—behave so badly.

Whether it spawns health or disease, the core code is stunningly compact. Converted into a digital file, the gigabyte or so of wetware stored in six tightly folded feet of the two strands of DNA that fused at the instant of your conception to create the original, single-celled *you* would provide about half an hour of snowy video on YouTube. Day by day, though, the snow is being translated into plain English, as new fragments of code are linked to the fuzzy medical vocabulary of the past, words such as *disease, mood,* and *behavior.* The Hatfields' endless feud with the McCoys? Blame the McCoys—dozens of them carry a rare gene that causes high blood pressure, racing hearts, adrenal gland tumors, and thus hair-trigger rage. Seven extra pounds? Point the finger at an obesity gene carried by one in six white Europeans. Nobody in your family gets cancer? Unusually good DNA-repair genes. All your relatives do? Hundreds of genetic links have been found, and new ones keep emerging. Alzheimer's? About a dozen genes and still counting. Can't sleep? It could be worse: a few dozen families on the planet carry the Macbeth gene, which murders sleep—they die young, of "fatal familial insomnia."

WETWARE DOESN'T TREAT clinically observable fatigue, weight loss, fever, anemia, and recurring infections—you have that cluster of symptoms because you have a rare form of leukemia, so you need Gleevec, which attacks an abnormal enzyme churned out in excess by the flawed gene associated with the "Philadelphia chromosome." Saquinavir doesn't set out to kill a whole virus; it attacks a key bond in a molecule that helps replicate HIV. Wetware science doesn't resign itself to the inevitability of side effects; it tracks them to the molecular variations that make some people allergic to aspirin, for example, and others unable to tolerate Camptosar, a powerful cancer drug. Pharmacogenomics matches the drug to specific genetic profiles found in some patients and not others. Gene therapy takes molecular medicine to its logical limit, using viruses to implant genetic fixes into the patient's own cells. And at the front end of it all, medical diagnosis becomes a microscopic trace of a sensor chemical embedded in a cheap sniffer—a dipstick that detects the pregnancy protein in the pee, or a chip-scale lab that detects cancer cells drifting in the blood. The frogs have been sent back to the pond.

The search for new wetware sometimes uncovers surprisingly powerful simplicity. A muddle of genes and lifestyles causes the high cholesterol that causes a raft of health problems. Lipitor ignores the muddle and just tweaks the body's cholesterol-production process. The flawed gene that apparently launched the adrenal-inflaming chemistry that enraged the McCoys also causes tumors in the eyes, ears, pancreas, kidney, brain, and spine, by interfering with production of cancer-suppressing proteins. Drugs are being developed to provide what's missing, and gene therapies that fix the body's own production line have shown promise.

The decoding of human bodies also reveals how the tiniest fragments of code can shape addictions, impulsiveness, and other complex patterns of behavior, which lead in turn to all sorts of other health problems once relegated to treatment by hapless psychiatrists. Fertile women are attracted to men with clusters of immune-system genes very different from their own, presumably because merging the two sets will produce kids who can beat more germs; women know who those men are because they smell right. There are genes for fidelity and promiscuity. Six related children in northern Pakistan chewed off their tongues, broke their bones, and burned and mutilated themselves—without feeling it. One killed himself jumping off a house on his fourteenth birthday. They shared a genetic defect that disrupts the body's production of a protein involved in transmitting pain signals. All their other sensory signals were transmitted just fine.

By assembling reams of fragmentary data to map out genomes and dredging genealogical databases to expose links between genes and disease, powerful computers play a large role in working out what drugs might target. Software developed to explain the dynamics of social networks is now being used to analyze biochemical networks inside the human body. By comparing genes in microbes, fruit flies, and humans, computers work out how certain molecules that evolved early in the history of life on earth became hubs with multiple links to functions and pathologies that developed much later in organisms like us.

Drug designers then unleash two astonishingly brilliant and powerful tools for designing magic-bullet molecules to strike the chosen targets. One enlists digital code to mimic life's. Seventy years ago, Alan Turing grasped that code is very good at decoding other codes and designed a machine that could decipher the Nazis' Enigma code. Today's computers are well on their way to being able to decode Andy Grove, or at least the parts that ail him, and then help design the right antidotes. Alternatively, magic bullets are created using the carbon-based biological code in laboratory animals. Either way, the vital core of medicine is now on the same plummeting-cost trajectory as chips and software.

———

THE ADVANCE OF molecular medicine still hinges, however, on the development of a vast new body of knowledge of a kind that Bill and Steve never had to worry about, because they had Andy at Intel. In the digital world, the hardware and software evolved in tandem, with chip designers and software programmers cooperating every step of the way. Drug designers, by contrast, have arrived on the scene billions of years behind nature. Nature neglected to provide manuals that explain how all the slivers of molecular code that it has created fit together and interact. When it finally got around to designing us, nature devised a very complex system. As we shall see, it had good reasons to do so, but yesterday's reasons don't help today's drug designers. Human bodies run on what digerati revile as spaghetti code: redundant webs of molecular synergies, compromises, contradictions, and barely suppressed antagonisms. The code-bearing molecules cooperate miraculously well at their best, and attack each other viciously at their worst.

No two bodies are quite the same, and the biochemistry of every unique body keeps changing throughout the course of life. The germs that assail us keep reinventing themselves. Human genes get remixed every time a new child is conceived. Thousands of variations in tens of thousands of building blocks create limitless possibilities for variation in how our bodies perform. And our bodies both shape and are shaped by the external environment, including the many microscopic forms of life that surround, colonize, and sometimes devour us.

Our diseases find ways to exploit this complex, resilient biochemical diversity. A few of them—the cholera bacterium, for example, or cardiovascular disorders caused by high cholesterol—have a single, quite simple vulnerability and can therefore be neutralized by a single magic-bullet drug. But most don't. And as medicine solves the easy problems, it increasingly finds itself confronting disorders that mirror the complex diversity that normally keeps us healthy. Tweaking bits of code to fix a problem here will often have unexpected consequences over there.

Pinning down exactly how all the pieces interact will be difficult. Impossible, say the critics of the "naive medical reductionism" on display earlier in this preface. It certainly was impossible until very recently. But with the vast and ever-expanding power of digital technology now at our disposal, science can and will work out enough to let us take control of a great deal.

The search for a new drug is, increasingly, a very expensive search for information about how human bodies function. That search accounts for a steadily rising fraction of the front-end cost of most drugs. Almost all of the new information that is gleaned from the clinical trials of a fundamentally new drug is needed to ascertain, directly or indirectly, how human bodies operate at the molecular level. That learning process inevitably continues after a drug is licensed and comes to be used far more widely in the marketplace. It will be repeated again and again, with one drug after the next. The information acquired will end up in massive, detailed, and very valuable databases that will power analytical engines that will expose the architectures and dynamics of countless molecular chain reactions and networks that make human bodies function well or badly. If that sounds far-fetched, suffice it to say for now that medicine already worked out the biochemical details and built the huge databases for at least one retrovirus. Without them, we never would have managed to design the drugs and assemble the cocktails that we currently rely on to control the multiple different strains of HIV. Similar databases and analytical engines are now being used to guide the treatment of various cancers. A human body is far more complex than a retrovirus, but the power of the technology now engaged in decoding life is doubling every year, if not faster.

CODE HAS THE tendency to rise from servant to master to sole proprietor. The first autopilot kept a plane cruising at steady speed and altitude; the cockpit of the future, it is said, will have a pilot, a computer, and a dog, the dog there to bite the pilot if he touches the computer. Much of medicine is now on a similar glide path.

It touches down in the office of the doc-in-a-box at Walmart. The largest retailer in the world is now set on "becoming the largest provider of primary healthcare services in the nation," monitoring complex diseases, dispensing powerful drugs, and "dramatically" lowering the cost of health care. There is good reason to believe that Walmart can deliver. Our health care system depends on two fundamentally different types of care—call them hands-on medicine and molecular medicine—which are propelled by fundamentally different economics. The people who run Walmart are quite smart enough to focus on the one that they can probably manage better than anyone else.

Doctors, hospitals, beds, and the rest of hands-on medicine still account for seven out of every eight dollars that we spend on medical care. These services are expensive, and costs rise in lockstep with the number of patients treated. But by the time a molecular seed of a problem has morphed into a fever, clogged artery, or lump that can be soothed or cut out by human hands, the problem is far worse, much more expensive to treat, and often unbeatable. And when it gets to that point, the biggest cost of health care is the routinely overlooked, off-budget cost of adults in their prime taking care of sick kids and aging parents, or lying in bed themselves, unable to work.

Molecular medicine plays by completely different economic rules. Developing the first pill and its user manual is very expensive. But, as with software, almost all the value of a new drug is know-how—information stored in the drug's design, and in the databases that describe how it will interact with the different biochemical ecosystems in which it may land. Once found, the know-how can be replicated and shared very cheaply, forever. And while the cost of this know-how spirals down, its power and value spiral up. Tomorrow's know-how builds on today's.

When it gets good enough to routinize the doctor's role, it gets as cheap as Walmart, displacing the doctor almost completely and thus dramatically lowering the cost of care. Vaccines, antibiotics, and the first handful of drugs developed to control a still short list of problems rooted in our chemistry have saved us far more money than we spent developing them. Take your Lipitor, fire a heart surgeon.

Walmart will be able to provide excellent, cheap care because the doc-in-a-box has so much powerful wetware within easy reach—the biochemical sniffers, dipsticks, and assays that perform most of the diagnosis, and drugs that can be used quite well for by-the-book treatment. Yesterday's brand-new medicine is today the cheap medicine of free riders coasting on someone else's genius and capital. Making yesterday's medicine widely available is a good thing. Walmart—or other companies like it—will do it very well. About three-quarters of current U.S. prescriptions are for generics. Walmart already offers a thirty-day supply of hundreds of them for $4.

But because so much of the cost of wetware is incurred up front and must be paid for by small groups of patients who use new drugs before their patents expire, wetware's power to slash health care costs isn't immediately visible. Information markets thrive only under the control of people who know how to distinguish capital investment in tomorrow's know-how from the economies of free riding on yesterday's, and who are determined to accumulate capital rather than debt.

———————

THE POLICIES AND rules that currently control the science and economics of molecular medicine now stand as a monument to medicine's ignorant past. They were cobbled together at a time when nobody could even read the molecular-scale code that controls health and disease. Drugs were designed mainly by hunches and guesswork, and very few worked well. The safe and effective use of drugs depended on gathering purely statistical information about how they affect high-level clinical symptoms. Governments could compile and make sense of the crude statistics better than anyone else and thus were in the best position to decide what we needed and how much it was worth. The system assumed broad areas of biochemical uniformity, conflated differences, and steered medicine relentlessly toward generic drugs for generic patients.

To implement the 2010 Affordable Care Act Washington, by all current indications, plans to do more of the same—much more. Washington first required drug prescriptions (for narcotics) in 1914. It began seriously regulating drug chemistry in 1938. In 1951, it empowered the Food and Drug Administration (FDA) to require prescriptions at its discretion. Federal drug law also defines diagnostic biochemical sniffers as "drugs," and the FDA decides which may be sold and to whom. Now Washington plans to dictate how doctors diagnose and prescribe. Which means we are now poised to witness the most audacious attempt to take control of the flow of information that the world has ever seen. At stake is how America will read, decipher, and master the code stored in the largest, most valuable, and most important repository of information on the planet—humanity itself.

Washington will appoint the experts who serve on the panels that write the scripts that determine which diagnostic tests are performed and which drugs, if any, are then administered. The scripts will determine which bits of biochemical code will—or won't—be scrutinized in different patients and which aspects of the patient's response to treatment will then be tracked. After doctors diagnose, prescribe, and track on script, they will report to Washington how things worked out. Washington will decide what it all means, and then write more scripts. Current law already places strict limits on how drug companies may communicate with doctors. The people involved at the two critical stages where all the action begins and ends are thus barred from collaborating in any way other than under Washington's strict supervision, according to rigid protocols Washington approves in advance. Patients, one must concede, will remain free to blog about their own experiences with a drug. So will academics, paymasters, statisticians, newspaper reporters, and politicians of every stripe, so long as they don't treat live patients. Washington, in short, now aspires to catch up with Europe and Canada in making Walmart's doc-in-a-box medicine universally available, cheap, and managed from the banks of the Potomac. This, we are told, will deliver better medicine to all, at lower cost.

It will do the opposite. Walmart can't deliver tomorrow's cures, but it doesn't obstruct their development and doesn't wish to. To get the really cheap medicine, Walmart has to wait for private capital to develop it and for the arrival of cheap generic drugs when patents expire. Government paymasters can ride those coattails, too, at the same pace and price, but why wait? Washington can't afford the drugs that are still under patent, but by positioning itself as the overwhelmingly dominant buyer it acquires the bargaining power to devalue patents and get to really cheap much sooner. And by doing that, Washington drives away the private capital that develops the drugs to subdue the diseases that are still ineffectually treated at great expense by the helpless doctors practicing hands-on medicine.

Pushed to the limit, Walmart medicine supplied by Washington leads to crony medical capitalism. Drug companies accept the fact that their real customers are the government's paymasters, and behave accordingly, focusing their attention on the diagnostic tools and drugs that Washington is willing to pay for. Doctors are conscripted into the ranks of the official stakeholders and junior partners, diagnosing and prescribing drugs as dictated by Washington's scripts. Patients, in whose bodies lies almost all the information that ultimately determines the safety and efficacy of treatment, do as they're told.

This is also a process doomed to sink ever deeper into bankruptcy. Cheap care, universally available, is the most expensive option when it delivers all the compassionate—but ineffectual—doctor-centered caring that anyone could ask for. The molecular medicine that works ends up cheap, but it starts out very expensive. Most of its front-end cost is investment in the most valuable and durable form of capital—pure biological science and technical know-how. As it accumulates, this kind of capital continues to make us richer, not poorer. But public health authorities who can't afford to invest in the medical future have little incentive to reform public policy in ways that will accelerate its arrival.

———————

THE TECHNOLOGIES THAT abruptly transformed medicine into an information industry came of age as yesterday's crowd-based medicine was reaching its limits. It had beaten cholera, smallpox, and other one-size-fits-all infectious scourges of the past, but its tools couldn't handle the more complex diseases that then surfaced.

Here's the rest of the story in three sentences. We can now see, and can probably find ways to control, almost any of the molecules of life. The hard part is reassembling the pieces—working out the connections between what we can see and probably control down there with what we wish to accomplish up here. We need policies that will allow biochemists, doctors, and patients to collaborate in ways that will unravel how human bodies work without the process taking so long and costing so much that it never gets started at all. (For outlines of what such policies might look like, see Chapters 5, 11, and 16.)

The book as a whole is structured around the nature and power of the new science and how it will transform medicine beyond recognition—if we let it. The focus throughout is on that big *if*—how the science itself, and the process by which it is transformed into useful medicine, collides with the regulatory status quo, and what changes are needed to clear the way.

The objective in the first part of the book is to describe the essence of the molecular science revolution that is well under way, and the process that translates the science into applied medicine. How do we see, understand, and systematically control health and disease? In brief, not Washington's way—not anymore. Medicine is coming apart. The medicine of crowds has been overtaken by the medicine of molecules, which exposes and exploits the common biochemical factors that separate and unite groups of patients and diseases in ways never previously discerned or imagined.

Part Two addresses how Washington judges and certifies the quality of molecular medicine, how it briefly unleashed the full power of the new science during the early years of our war against HIV, and how its relapse to the past now suppresses the efficient, orderly development of that science. How should we regulate the process that develops and certifies the new science? In brief, the venerable myth of the FDA-scripted "gold standard" for drug science is fading fast. Most of what we still call the "drug licensing" process now hinges on licensing patients to fit our drugs. Washington spent forty years creating protocols for the development of crowd-science medicine, and it has spent the last thirty wondering how to fit molecules into those protocols. The system is currently frozen in the headlights. It can't handle the complexity and torrents of data that now propel the advance of molecular medicine. It isn't just lagging behind the new science; it has fallen so far behind that it now obstructs it at almost every turn. The challenge for the FDA, ironically, is how to regulate the science, now that there is, for the first time in medical history, a true, rigorous, mechanistic science to regulate.

Part Three addresses the parallel universe of the dollar doctors. Who pays for what, when, and how does that determine what we get and don't get? In brief, medicine that was centrally managed and funded made excellent sense during humanity's long war against the old infectious scourges; it is now stifling the development of far more powerful molecular medicines and the capture of even greater economies. Today's heavy-handed, top-down pursuit of what bankrupt governments call affordable health care, equally available to all, can't be squared with the dynamic, innovative, molecule-by-molecule development of cures that work. While their development requires huge up-front investments, those cures are the ones that end up cheaply available to all and displace much more expensive doctors when they reach the autopilot stage. One-size-fits-all medicine managed and funded from the top would lock us into the old world of helpless care that keeps getting more expensive. In every area where the power of molecular medicine has been unleashed, the cost of health care is falling fast. To keep private capital engaged in the pursuit of ever more complex diseases, we will have to change the economic rules as fundamentally as biochemistry has changed the medicine.

Finally, how do we bridge the gulf that now separates America's bipolar culture of death and life? In brief, we stop dwelling on the rights that Washington has spent much of the last fifty years inventing and expanding, and rediscover the rights and responsibilities that were valued and affirmed by our grandparents.

While science has dramatically expanded its power to control our biochemistry, we have lost the social capacity to unleash that power against the diseases that are now killing us. Washington boldly affirms our individual rights to control our own bodies and mess with our own biochemistry pretty much as we please—and also receive all the health care we then need. To make sure we get it, Washington will minutely regulate the development and use of diagnostic sniffers and drugs, pay for them if we can't, and flex its muscle to limit how much it has to pay. This is the culture of death.

The drugs of the future will emerge only from a flexible, adaptable process of discovery that nurtures communication and collaboration among investors, biochemists, insurers, doctors, and patients—a process that allows small, nimble investors and biotech companies to thrive alongside the established giants. The development of these drugs will require an economic environment that affirms property rights in the information involved, including rights that will spur the creation of the vast biochemical databases on which the development and effective use of new drugs now depends. It will require, above all, a public understanding of the power of molecular medicine that fosters a culture of confident, eager, optimistic engagement in the process of maintaining health, beating diseases, and extending life spans. Steve Jobs worked his magic by giving us insanely great things while simultaneously developing a culture that mobilized and coordinated sprawling, vibrant communities of creators, manufacturers, and consumers. Dynamic, information-based industries are dynamic because they are decentralized. They allow information to move freely. They are willing to take risks, accept failure, and richly reward success. Their employees have a strong sense that they are engaged in something important and valuable. Their customers are given the freedom, and also the responsibility, to learn how things work and to interact directly with vendors to help solve problems and propel the next round of innovation.

In markets that deal with the life-and-death stakes of diseases and drugs, none of this means dispensing with regulators altogether—far from it. It does, however, mean adopting policies that allow biochemists, doctors, and patients enough flexibility to confront and collaborate to master the complexities of biochemical reality.

———

AUTHORITIES COULDN'T CONTAIN silicon, software, Grove, Jobs, Gates, or Google in a state-regulated bottle. They won't do any better with code embedded in biochemical media. The private capital needed to develop the medical capital of the future will find its way to places that maintain a regulatory and economic environment that allows the money to develop the science, and it will develop drugs for markets that will allow the money to recover its costs and earn profits commensurate with the economic risks.

Whether the United States will lead or follow remains to be seen. Thirty years ago, it seemed fairly clear to techno-utopian determinists that nothing could stop the power of digital technology. Far less evident was whether America would act faster than other parts of the free world to clear away the regulatory obstacles that stood in its way. Dismantling the regulated-monopoly mentality of the telecom past to unleash the largely deregulated digital future took another two decades, with many detours, re-lapses, and unnecessary delays, multiple trips through both state and federal courts (including several to the U.S. Supreme Court), and the passage of a major new federal telecommunications law, but we got it done. America has led the world in the development and deployment of digital technology and has reaped enormous benefits as a result.

Once again America has to choose. We can embrace the extraordinary power of biochemical technology, implement policies to unleash the market forces that will propel the biochemical information revolution, and allow America not only to lead but to move far out ahead of the rest of the world—or we can cede to others control over the code of life, the most free, fecund, competitive, dynamic, and intelligent resource on the planet.

Part One

COMING APART

1

THE TRIUMPH–AND
LIMITS–OF SOCIALIZED MEDICINE

"THERE HAVE BEEN at work among us three great social agencies: the London City Mission; the novels of Mr. Dickens; the cholera." In *The Moral Imagination*, historian Gertrude Himmelfarb quotes this reductionist observation at the end of her chapter on Charles Dickens; her debt is to an English nonconformist minister, addressing his flock in 1853. It comes as no surprise to find the author of *Hard Times* and *Oliver Twist* honored in his own day alongside the City Mission, a movement founded to engage churches in aiding the poor. But what's *Vibrio cholerae* doing up there on the dais beside the Inimitable Boz?

It's being commended for the tens of millions of lives it's going to save. This vicious little bacterium has just launched the process of transforming ancient sanitary rituals and taboos into a new science of epidemiology. And that science will frame a massive—and ultimately successful—public effort to rid the city of a long list of infectious diseases.

Socialized medicine's finest hour arrived a century later, on October 16, 1975, by the marshes of Bhola Island, off the coast of Bangladesh. There, in the frame of three-year-old Rahima Banu, the World Health Organization finally cornered smallpox, the most efficient killer on the planet. Then as now, there was no known cure for the highly contagious disease, but vaccinating others on Bhola Island kept the virus from skipping to new human hosts, and little Rahima was the last one left.

In the 1960s, at the height of the Cold War, the Soviet Union and the United States had joined forces under the aegis of a United Nations affiliate to beat the disease. Over the course of the next two decades, they bought billions of doses of vaccine, deployed tens of thousands of workers, mobilized national armies to isolate infected regions, prohibited public meetings, quarantined hotels and apartment buildings, dispatched helicopters, airlifted refrigerators, sent in doctors and nurses, established rings of immunity around newly infected areas, and then tightened the rings until only one ring surrounding one little girl remained. And she, standing tiny but tall, finished off the most virulent strain of the virus on her own.

America underwrote much of the global war on smallpox and helped save hundreds of millions of lives worldwide, and because we no longer have to protect ourselves against smallpox, Americans today save as much every month in direct and indirect societal costs as their grandparents spent on the entire smallpox campaign. According to estimates made in 1994, every dollar spent on polio vaccine saved about five times as much in such costs. The measles immunization payoff was thirteen to one. In 2001, every dollar spent on immunization with five widely administered vaccines was saving an estimated $6 in direct medical costs and another $12 in costs of missed work, disability, and death. Socialized medicine's hundred-year war against germs did far more to improve human health and extend life expectancy than all the rest of medicine developed before or since.

But these triumphs of socialized medicine are behind us now. The medical future looks nothing like the medical past—the diseases are different, and so are the cures. While medical science moved on, Washington dug in. Our medical-regulatory complex—a briar patch of scientific and economic proscriptions, mandates, subsidies, and patents, plus one strange, little-known, but extremely important form of copyright—remains rooted in the scientific methods and public policies that coalesced between 1853 and the early 1980s. These policies, as outlined briefly in this chapter and discussed throughout the rest of the book, were designed to regulate ignorance, not knowledge—the dearth of molecular medical science, not the science itself, nor its efficient, orderly development.

———

POUND FOR POUND, bacteria, viruses, and other microbes contain far more intelligence than we do. They are the closest nature gets to pure code. They are also the most ancient, nimble, creative, and persistent developers of new code on the planet. They have survived planetary catastrophes ten thousand times longer than creatures like us have walked the earth. They probably outweigh all the rest of life combined. They thrive under the arctic permafrost, alongside thermal vents at the bottom of the ocean, under miles of rock, and most everywhere else on or anywhere near the earth's surface. Our own bodies host an estimated ten germs for every human cell.

A fistful of germs that specialize in dining on people evolved over the course of three thousand years of urban human history. These legacies of the distant medical past developed a molecular peg that fits neatly into some molecular hole that many of us share. Quite a number of them also found some ingenious way to provoke the sneeze, the violent diarrhea, or even the lustful itch that helps the germ spread. The cholera bacterium, for example, persuades cells in your gut that you've eaten a dangerous amount of salt, which must be flushed out with water sucked from your body.

Germs discovered the joys of socialism long before we did, and in health matters made communists of us all. An epidemic—from the Greek meaning "upon the people"—was the democracy of rich and poor incinerated indiscriminately by the same fever, or dying indistinguishably in puddles of their own waste. The crowded city served as septic womb, colony, and mortuary.

Smallpox was the Mao of microbes—it killed three hundred million people in the twentieth century alone. Sometimes called the first urban virus, it probably jumped from animals to humans in Egypt, Mesopotamia, or the Indus River Valley at about the same time that the rise of agriculture began drawing people together in towns and cities. Smallpox has also been called nature's cruelest antidote to human vanity. Princes broke out in the same pustules as paupers, reeked as foully of rotting flesh, and oozed the same blackish blood from all their orifices. Alongside millions of nameless dead lie kings of France and Spain, queens of England and Sweden, one Austrian and two Japanese emperors, and a czar of Russia.

When monarchs were dying alongside peasants, the language and politics of health and disease honestly tracked medical reality. The main threats to the "public health" threatened everyone. They also pointed to a compelling need for collective solutions. Disease was caused by invisible, external agents that individuals couldn't control on their own. We were all in the same us-against-it battle to the death. (Though the arrival of the invisible *it* was often blamed on foreigners—syphilis was the "French disease" for the English, "la maladie anglaise" for the French, the "Polish disease" for the Russians, and the "disease of the Christians" for the Arabs.)

For Dickens, the filth in the Thames River symbolized London's insidious taint, its ubiquitous, effluvial corruption. The urban pathologies he described in *Our Mutual Friend* in 1864 were as familiar to Londoners as the river. What social historians sometimes fail to note, however, is that here art imitated life. By the time Dickens was placing the Thames at the center of London's many ills, a new science—epidemiology—had already emerged to move the river far beyond metaphor.

———————

THE SYSTEMATIC CRACKING of microbial code began in a city periodically assailed by what the *Times* of London would call the "Great Stink." It began rising from the sewage in the Thames during the hot, dry spring of 1858. By the middle of June, the Parliament buildings, which overlook the river, were uninhabitable. "A few members, bent upon investigating the matter to its very depth, ventured into the library but they were instantaneously driven to retreat, each man with a handkerchief to his nose," the paper reported. At the Court of Queen's Bench, a surgeon declared that the smell was endangering jurymen, counsel, and witnesses, asserting that "it would produce malaria and perhaps typhus fever."

This "miasma" theory of disease was widely accepted throughout much of the nineteenth century. Most diseases—perhaps even all—were thought to be transmitted mainly by "bad air," that is, air corrupted by decaying corpses, the breath of the sick, sewage, rotting vegetation, or whatever else might be producing a vile smell.

Epidemiology, the medical science of crowds, was born when physician William Farr was appointed controller of London's General Register Office in 1838. Directed to find out what could be done to defeat cholera, Farr began systematically recording who was dying and where. What mattered, he discovered, was at what elevation above the Thames you lived—higher was safer. The stink was indeed lethal, he concluded.

Another fifteen years would pass before another English doctor, anesthetist John Snow, figured out what was really occurring. London's gravity-driven sewers emptied into the Thames, so the farther down-sewer you lived, the more likely you were to draw foul drinking water from the local well. A year later, after a particularly nasty outbreak of cholera in Soho, Snow saved countless lives by persuading parish authorities to remove the handle from the neighborhood's pump. Farr would acknowledge his error a decade later.

Farr and Snow had discovered a single key fact about the bacterium's code: cholera propels itself from my guts to yours through the water you drink. By pinning down the pathway of contagion, epidemiology transformed what had been, to that point, a long and fruitless struggle against a devastating public disease into a routine exercise in civil engineering.

All the legacy germs use public transit—the common water or air, most typically, along with rats and insects. Victorians would go on to squeeze a tremendous amount of public health out of their first, distant glimpse of this one tiny shard of hard science. In 1858, Parliament passed legislation, proposed by then chancellor of the exchequer Benjamin Disraeli, to finance new drains. The 1866 cholera epidemic was London's last. In 1872, Disraeli rallied his Tory Party around what his Liberal opponents derided as a "policy of sewage"— reforms involving housing, sanitation, factory conditions, food, and the water supply—and while he served as prime minister, these policies became law.

Until well into the twentieth century, in the United States as in Britain, public health depended on city bureaucrats above all. They wasted little time with sick patients, other than sometimes ordering them to lock their doors and die alone. The authorities focused instead on eradicating germs before they reached the patient, and that meant attending to the water, sewage, trash, and rats. In a recent survey by the *British Medical Journal*, public sanitation was voted the most important medical advance since the magazine was established in 1840. If we don't think of public sanitation as "medical" anymore, it's only because the municipal bureaucrats who followed Farr cleaned things up so well.

———————

WHILE THEY WORE out their welcome in public spaces, microbes could still thrive in the relatively private space of the public's lungs, fluids, and intestines. The pursuit of germs into the flesh of patients didn't really begin until the end of the nineteenth century. Edward Jenner's smallpox vaccine was already a century old, but it owed its existence to the lucky fact that the human pox had a weak cousin that infected cows. (We give our kids the "cow treatment"—*vacca* is Latin for "cow"—every time we do as Jenner did and challenge their immune systems with a vaccine made from a corpse, cousin, or fragment of a horrible microbe.) The production of other vaccines had to await the arrival of Louis Pasteur, Robert Koch, and the procedures they would develop to isolate microbes and then cripple or kill them.

While vaccines banish germs from individuals' bodies, health authorities quickly recognized that they are quintessentially public drugs. By interrupting the germ's chain of transmission, vaccinating enough people creates "herd immunity" that protects the rest. Vaccines are also cheap—far cheaper than the care provided by doctors and hospitals—and get cheaper when more people use them. Developing the front-end know-how is the expensive part.

Government agencies began buying and distributing vaccines, and never looked back. The U.S. Army vaccinated soldiers against smallpox during the War of 1812. During an 1815 epidemic, New York helped fund free vaccination for the poor, as did other cities and towns throughout the nineteenth century. Quality control slipped into Washington along with the money. In 1813, Washington appointed the army's supplier to serve as the nation's "vaccine agent," in charge of providing "genuine vaccine matter" upon request.

New laws, vigorously enforced, drafted the healthy public into the war on germs. England mandated universal smallpox vaccination in 1853. In February 1902, facing a smallpox outbreak, Cambridge, Massachusetts, set up free vaccination centers, mandated vaccination for all, and appointed a physician to enforce the measure. One resident, Henning Jacobson, refused to comply, claiming a constitutional right "to care for his own body and health in such way as to him seems best." But the U.S. Supreme Court upheld the "power of a local community to protect itself against an epidemic threatening the safety of all." Later state and federal enactments would require children to be vaccinated before they could attend public schools. Adults who traveled abroad had to be vaccinated if they planned to come home. Albert Sabin's polio vaccine would take things even further—the vaccine itself was contagious. A child swallowed a sugar cube soaked with live but weakened virus, and then went home and vaccinated his siblings and parents, too.

The germ killers didn't really get into the business of curing sick patients until the development of sulfa antibiotics in the 1930s, followed by penicillin and other antibiotics after World War II. Even then, much of the cure often lay in preventing the contagious infection—tuberculosis, for example—from spreading to others.

Year by year, governments expanded their control of the development, composition, distribution, and price of the germ-killing munitions. Launched with the help of the president whose likeness now adorns the coin, the March of Dimes funded the development of Jonas Salk's polio vaccine and oversaw the first large field trials. Soon after, Washington set up a program to promote and subsidize the vaccination of children nationwide.

Until World War II, vaccines and antibiotics had been developed and commercialized mainly by private initiative; government agencies would then begin buying and distributing them, confident that the private companies would continue to develop and market new vaccines however far the regulation, subsidy, and government purchases might be pushed. After the war, as the global smallpox campaign would demonstrate, many people in positions of influence concluded that a complete socialization of the war against germs made sense, and indeed it did—at least for the types of germs that they were dealing with at that time. The public authorities attacked the germs with genocidal determination, expelling them, one by one, from human society. Defiled by monstrous human fratricide, the first seven decades of the twentieth century were also the triumphant decades of public health. Within a decade after Jonas Salk announced his polio vaccine to the press in April 1955, the war seemed all but over.

———

ATAVISTIC MEMORIES OF diseases such as smallpox and cholera still shape the way we think. We can't help but feel that disease is an attack from the outside. We cling to the notion that simple, clear lines separate disease from health, and one disease from another. The attack is sudden, quick, and surprising. A discrete cause produces a discrete set of clinical symptoms—

fluxes, fevers, lesions, or lumps—that uniquely define the disease. By the time we realize we're sick, we're often in deep trouble. A cure proves its worth by making the nasty symptoms disappear without producing some other effect that's as bad or worse.

This simple view of disease was quite adequate during the long war against the legacy germs. The medicine of crowds was invented during this period. The foundations for all major political schemes to manage health care from the top down and cover most of its cost were put in place to exploit the limited tools that medicine developed to beat the great infectious killers of the past. At the time, a future of runaway demand for medicines to treat the diseases that would surface as the legacy germs receded was scarcely imaginable.

It was during this same century of germicidal warfare that Washington erected a huge, mind-numbingly complex edifice for certifying the safety and efficacy of all the medicine bottled by drug companies. A first statutory cornerstone was put in place in 1902, following a tragedy involving contaminated batches of a diphtheria antitoxin. A second, in 1938, followed a tragedy involving an antibiotic that had been dissolved in antifreeze to make it taste better. Important amendments addressing vaccines and antibiotics were enacted at the end of World War II. Between 1962 and 1981, driven largely by memories of the thalidomide disaster that had caused thousands of tragic birth defects in countries (the United States not among them) where the sedative was licensed and sold, Washington consolidated this authority and spelled out in microscopic detail what kind of science would be required to get a new drug to market.

A drug's effects would be judged by the effects it produces up here, where patients ache and worry, and where hands-on clinicians diagnose and treat—not down at the molecular level, where, say, tetracycline (we now know) latches on to a specific receptor on the surface of the cholera bacterium. A drug was expected and required to have pretty much the same effect in every patient suffering from the same, clinically defined disease, because medical science lacked any way to distinguish patients whom the drug would help from those it would harm.

To this day, Washington almost always requires and relies on the same kind of evidence—a side-by-side, statistical comparison of the health of two crowds—to decide whether a drug can do to some disease what better sewers did to cholera. Typically, one crowd gets the real thing, the other a placebo (commonly though incorrectly referred to as a sugar pill); when a reasonably good treatment is already available, the comparison may instead be drug versus drug. Independent doctors track clinical symptoms. The newly healthy and the still sick, the living and the dead, collectively vote the drug up or down.

In their day, these policies made good sense. The legacy germs so dominated human health and mortality that slow, subtle, selective killers were scarcely visible and didn't much matter. And medicine, in any event, had no idea how to treat them.

––––––––

SOCIALIZED MEDICINE TOOK care of the easy part. The mission was politically easy—the problems were big, visible, and horrible, and they threatened everyone. Compared with what would follow, the targets were huge and stupid—the dodo birds of microbes, destined for extinction soon after medicine discovered that they existed. For top-down managers of the science and economics of health, smallpox and Jenner's vaccine were as good as it will ever get.

As the legacy germs receded, they gave way to diseases—most rooted in our own complex, diverse human chemistry, some in previously unknown types of microbes—that are so gentle, subtle, slow, and complex that we often have a great deal of trouble recognizing them until it's too late. They dodge our immune system. They lie low for years, some from the instant of conception; then they often kill quickly. They force us to view and treat disease in completely new terms, not as the simple symptoms of a simple cause but as complex biochemical systems, sequences, and webs of causes and effects. The chemistry that gets hijacked to propel the disease often plays an equally critical role in maintaining good health. Any drug that targets it may cause a raft of side effects.

These diseases disassemble our health, fragment treatment, and pull us apart. They force us to confront our cultural discomfort with seeming to blame the sick, or focusing on differences tied to ancestry. Conditioned as we are to celebrate our differences, we still find it very difficult to discuss unhealthy lifestyles and flawed genes. But the relentless advance of molecular science is outing them all, regardless. Medicine's principal mission today is to provide antidotes to the unhealthy side of human diversity, diversity embedded in our own fissiparous chemistry. And as science has tracked these diseases down to their molecular roots, it has exposed the fundamental weakness of the crowd-based medicine of the past.

Statistical correlations that link clinical symptoms to suspect causes are what science uses to pluck the most primitive form of medical understanding out of the depths of ignorance. Victorian doctors didn't even need statisticians to connect boys who had a rare scrotal cancer to their days spent crawling naked through sooty chimneys, or mad hatters to some brain-rotting agent (mercury, we now know) in their hat shops. But even connections so clear that they jump right out of the numbers can't jump until somebody thinks to add "chimney," "hatter," "scrotal," and "loony" fields to patient records, and then assembles a database large enough to include a good number of check marks in these weird little boxes. And statistics only obscure what matters when the common clinical symptoms are produced by diverse and complex arrays of underlying molecular causes. Bad-diet statistics have been hard to pin down, because such things as cancer, heart attacks, and strokes are quite common among the elderly in any event.

And the statistical analysis of high-level clinical symptoms always starts later than it might and often ends much too late. In the 1960s and 1970s, at a time when the human immunodeficiency virus (HIV) was spreading unnoticed across the country, a clotting factor used by hemophiliacs was being extracted from human blood plasma collected from multiple donors and pooled. Many thousands of hemophiliacs died because HIV takes years to turn into acquired immune deficiency syndrome (AIDS), and crowd doctors didn't start worrying until clusters of AIDS patients surfaced. Testing the first HIV drugs presented exactly the same problem. The studies that Washington had traditionally required to prove that drugs work can't be completed any faster than diseases typically progress.

Science no longer needs to wait for high-level clinical symptoms to surface; it now knows how to study causes and effects down at the bottom. Bad statisticians, however, can always find ways to lump back together what good science takes apart. When they work for the government, political and economic objectives often force them to do so. To pick just one example among the many (less inflammatory) examples discussed throughout the rest of this book, it is easy to cobble together statistics to suggest that contraception and abortion are—or are not—linked to higher rates of breast and cervical cancer (see Chapter 4).

But, good or bad, statisticians must now compete with modern medicine's most powerful antidote to the health care policies of the Great Stink past: the sniffing technologies that allow medicine to read and track—and then develop antidotes to treat—molecules and cells, rather than crowds.

MEDICINE'S RECENTLY ACQUIRED ability to read every letter of the code of life has added a short, rightward-leaning codicil to the smallpox story, a familiar account of the discovery that led to socialized medicine's finest hour and which surely still warms every left-leaning political heart. On May 14, 1796, Jenner injected eight-year-old James Phipps with cowpox pus taken from lesions on the hand of milkmaid Sarah Nelmes, and found that this mild infection protected Phipps from deliberate attempts to infect him with an aged and thus weakened form of the human pox. Jenner published his findings two years later, content to gift the most valuable pharmaceutical discovery of all time to suffering humanity. "Yours is the comfortable reflection that mankind can never forget that you have lived," wrote Thomas Jefferson in a letter sent to Jenner in 1806.

But as vaccinators ramped up their attack, the smallpox virus fought back. The strain of *Variola major* that emerged three thousand years ago killed about 30 percent of its victims. A milder strain that appeared in the late nineteenth century killed a mere 1 percent. A 12 percent killer surfaced in 1963. The first and worst strain perished on Bhola Island in 1975, but wiping out its siblings took another two years. To this day, as David Koplow recounts in his 2003 book *Smallpox*, "no one knows where, when, or how these less noxious smallpox relatives crept into existence."

We do know, however, that the vaccine that ended up beating them all wasn't Jenner's. The details are lost in history and Koplow himself doesn't speculate about them, but it's easy to surmise how this vaccine came into being.

Picture how the market for what began as Jenner's vaccine operated through all but the last few decades of its two-century run. Infectious muck was scraped from scabs found on cows or milkmaids, then scraped back into human arms. It was transported on sailing ships to America by moving it from arm to arm, a human chain letter. Unwashed human hands, knives, and needles did the scraping, inevitably picking up more muck along the way, including smallpox itself. Countless unregulated purveyors of vaccine got involved, many of them careless, incompetent, or worse. Washington did its bit here, too—the national "vaccine agent" it appointed in 1813 accidentally sent real smallpox instead of vaccine to North Carolina, infecting seventy people and killing ten.

But people apparently noticed and spread the word: *This shot works better than that one*. Choices encouraged this way and accumulating over the years had bred better grains, dogs, and sheep, and now they set about breeding a better vaccine. And on it went, until the huge Wyeth Labs picked what it considered the best of the breed, got it licensed, and—with a little help from the left—obliterated the greatest bioterrorist of them all.

The story of *Variola major* ended in 1975, but *vaccinia*'s didn't end until a quarter century later, when gene sequencers found out what it really was. The several strains of smallpox in the wild, we now know, were all eradicated by just one virus—the same "novel, separate creature"—in all the needles. It appears to be a remix of cowpox, another cousin that poxes horses, and the human pox. It was created, as Koplow notes, by means that were "somehow inadvertent, invisible to the practitioners, and global." Or as a biologist and an economist whose lives overlapped Jenner's might have put it, by means of natural selection and the invisible hand.

2

SNIFFERS

In EARLY 2012, scientists at Stanford University described how they had spent the previous two years tracking DNA, RNA, cell proteins, antibodies, metabolites, and molecular signals—some forty thousand biomarkers that yielded many billions of bits of data—in the body of geneticist Michael Snyder, the team's senior member, to create the first-ever "integrative Personal 'Omics' Profile": an "iPOP." Though Snyder had no family history or conventional risk factors, the data revealed a genetic predisposition to type 2 diabetes, and then, later in the study, tracked the onset of the disease in what has been described as "the first eyewitness account—viewed on a molecular level—of the birth of a disease that affects millions of Americans." Then the iPOP team watched the diabetes markers revert to their normal state in response to treatment. And that, as discussed shortly, was only the beginning.

The diagnostic power that delivered Snyder's iPOP, and considerably more, will soon be embedded in chip-sized labs and mass produced at low cost. "You should be able to get 5,000 tests done with one drop [of blood] so we can get a better idea of what's going on in your body," says Snyder. "I think people who are at risk for certain diseases could do a simple home test. You could probably monitor yourself every month so you can catch diseases early."

The iPOP, as performed by the ever more affordable, compact, and powerful molecule-sniffing technology, is where the medical search for causes and effects ends. In cholera-plagued London, Farr and Snow saw nothing much smaller than a river and a neighborhood water pump and counted bodies, living and dead. Until very recently, standard lab tests included in an annual health check measured a few dozen variables. Now, gigabytes of molecular data can cascade from a single body into a powerful computer to expose the elemental logic of the most complex causal chains and networks that maintain health, or spawn and propel a disease, or subdue it.

The power to read the code that choreographs life is transforming medical science in four fundamental ways. By exposing the complex biochemical diversity that often lurks underneath a single set of clinical symptoms, it disassembles and redefines diseases. By revealing precisely how diseases are spawned and propelled down at the molecular level, it reveals what drugs should target, and the targets then serve as templates for designing the drugs. By revealing why a drug performs well in some patients but not others, it provides precise biochemical criteria for selecting the patients who should use a drug. And by transforming the definitions of diseases, the design of the drugs, and the manner in which they are prescribed, the power to read and understand the molecular code also transforms how Washington should regulate drug science and economics. Along with how doctors practice medicine, and how the rest us think about our health and how to maintain it.

———————

UNLIKE HUMAN BEINGS, any Alsatian in the dog pound can smell, at parts-per-trillion concentrations, biochemicals uniquely associated with low blood sugar levels in diabetics; epileptics on the threshold of a seizure; human lung, breast, or bladder cancer; dairy cows in heat (ready for artificial insemination); and things as varied as dynamite, cocaine, cell phones, bedbugs, mercury spills, weeds, toxic mold, and the polycarbonate of pirated DVDs.* The stunted human brain, however, is now developing cheap, compact sniffers that allow it to perform just as well, and even better.

The most familiar body sniffer is probably the one that scans urine for the protein that choreographs pregnancy, but there are oodles of other molecules in our bodies with important stories to tell. Other FDA-approved, home-use sniffers detect, among other things, glucose, cholesterol, and fertility proteins in blood; infectious microbes, nicotine, and illegal drugs in urine; and HIV antibodies in saliva. The diagnostic end of medicine—the purely informational end—is now on the same plummeting cost curve as the microprocessor.

The technologies for assembling and mass-producing the biochemical ingredients and choreographing the biochemical reactions that power the iPOP itself, and the many less-ambitious molecule sniffers already on the market,

————————

* Dogs have exceptional sampling systems, too. Bloodhounds use their huge, floppy ears to waft odor up from the surface of the ground into their probing nostrils. Other breeds use their tongues. In 2006, a beagle named Belle won a Wireless Samaritan Award for saving her diabetic master's life. She had been trained to periodically lick her owner's nose to check his blood sugar, and to autodial 911 on his cell phone if it dropped too low.

have all been mastered. Arrayed on chip-sized, micro-electromechanical laboratories, sniffers are now becoming complete bioscanners that can, for a few dollars a whiff, search a cheek swab or drop of blood for hundreds—and soon thousands—of genes, proteins, fats, and other biomarkers. Sensor chemicals on the surface of plastic or paper cards mounted in a breathalyzer can detect a hundred or so biomarkers, including those signaling the presence of lung cancer and tuberculosis.

More complex sequences of assays are now performed by huge, fully automated banks of compact diagnostic machines that can quickly and cheaply diagnose infections, genetic abnormalities, and biochemical imbalances of every kind, in as many specimens of urine, blood, saliva, or mucus as anyone cares to swab on a Q-tip, smear on a card, or dribble into a little cup. They can sense variations among our body's tens of trillions of cells, salient features of the trillions of microbes that we also host, and the biochemical soup that bathes all the rest.

Sniffers reveal within each individual human body a rich, intelligent, and intelligible text, a catalogue of descriptions and instructions composed by nature and nurture over the course of four billion years, with all the accidental and deliberate editing wrought by sex, love, instinct, accident, and conscious choice. To sniff in the ways sniffers now make possible is to converse lucidly with one's ancestors—and with the cholesterol in yesterday's lunch, and with the salmonella on the lettuce. To sniff brain chemistry (yes, the biotech wizards are working on that, too) is to read one's own thoughts, memories, emotions, and instincts.

Used in solitude, sniffers allow completely private scrutiny of the most compact, brilliant, and powerful script that any of us will ever have a chance to read. And the reading will launch all sorts of conversations, private and public, as well as social initiative and political action.

———

THE IPOPING OF Michael Snyder began when he was, by all clinical appearances, perfectly healthy, and it thus established a biochemical baseline for his personal clinical health. The early genetic scan, however, revealed a genetic propensity for high cholesterol, which he already knew about, and also for diabetes, which came as a surprise. He then watched his cholesterol level drop sharply when he started on a cholesterol drug. After his blood sugar levels suddenly jumped on day 301 of the tracking, he watched aspirin, ibuprofen, exercise, and a low-sugar diet wrestle them back down.

For Michael the patient, that might have been enough. But for Professor Snyder the scientist, there was more to learn. Analysis of the iPOP data also revealed how his RNA was activating different genes as his health changed during the course of the study. As the patient recounts, "We generated 2.67

billion individual reads of the [relevant RNA molecules], which gave us a degree of analysis that has never been achieved before. . . . This enabled us to see some very different processing and editing behaviors that no one had suspected. We also have two copies of each of our genes and we discovered they often behave differently during infection."

The researchers suspected a possible link between a viral infection and Snyder's blood sugar surge twelve days after its onset, and so they zeroed in on about two thousand genes that were fired up during that period, plus another two thousand that throttled down. They found some that help control insulin—links involving inflammatory proteins and antibodies, among them an autoantibody that targets a human insulin receptor. The data thus pointed to "unexpected relationships and pathways between viral infection and type-2 diabetes." As one of Snyder's colleagues notes, an analysis of this kind reveals how a patient's complex control systems interact with his or her own chemistry and the environment, and thus point to how medicine "can best target treatment for many other complex diseases at a truly personal level."

In the iPOP world, it isn't just the medicine that gets personal; the science does, too. The science that describes the biochemical structure and dynamics of the disease and determines the efficacy and safety of the antidotes still involves a comparison of two or more patients, but they have the same name. "In a study like this, you are your own best control," says Snyder. "You compare your altered, or infected, states with the values you see when you are healthy."

The development of this personal science does of course build on a large body of knowledge previously acquired from other patients, and the data gleaned from Snyder's body will surely be used to help refine how other diabetics use iPOP technology going forward. As Snyder notes, researchers with access to such data should be able to converge on a much smaller number of variables that can predict future blood sugar health and track the rise and fall of diabetes and other diseases.

But as Snyder also points out, the picture that will likely emerge from this bottom-up, data-extravagant science isn't likely to please the crowd doctors. There are probably "many reasons why someone is at risk" of type 2 diabetes. "Diabetes is really hundreds of diabetes, and they just have one common characteristic which is a high level of glucose," he says. Different patients therefore require different treatments. "Some respond to metformin [a drug that suppresses glucose production in the liver], some don't. Some respond to anti-inflammatory medicine, some don't." And with diabetes, as with many other diseases, the key to effective prevention or treatment is "to catch it earlier."

The treatment of many other complex diseases already depends on batteries of ongoing tests to monitor both intended and unintended effects of complex drug cocktails. Researchers at Massachusetts General Hospital recently developed a chip-scale sniffer of cells shed from primary tumors that circulate in the bloodstream—they call it a "liquid biopsy." Cell counts provide an immediate indication of whether the primary cancer is retreating or advancing; genetic analyses provide an early indication of the emergence of drug-resistant cells. Breast cancer, one form of diabetes, Alzheimer's, and other disorders leak protein signatures into our saliva, and detectors are being developed to biopsy spit instead of blood.

Doctors have long relied on informal case reports in medical journals to share what such tests may reveal about treatment regimens. They now have far more data to learn from, and are rapidly systematizing the process of pooling and analyzing the data they gather as they treat. Many patients already share data on their own terms with the people they choose, data that they would never willingly share on Washington's terms with Washington. One website has coordinated a social-network-based online trial of lithium treatments for Lou Gehrig's disease; another is compiling data from patients suffering from a rare gastrointestinal cancer; yet another recently identified five genetic associations for hypothyroidism from a comparison of 3,700 individuals with the disease from over 35,500 controls. Many other sites and services are coordinating similar ad hoc studies; they call it "crowd-sourced science." As HIV patients did so effectively (albeit without the help of the Web) in the 1980s, patients also use such sites to discuss what seems to work and what doesn't, lobby for more research, and demand faster access to new drugs.

And the self-sniffers are using what they find, along with all the guidance they can receive from other sources, to get out ahead of, or second-guess, the officially approved pronouncements about how drugs should be used. There are obvious risks in that second-guessing—and also one indisputable benefit: the only thing that matters to the patient is how the drug previously performed in biochemical ecosystems very much like his own. The official pronouncements often arrive late, straining to address what's best, on average, for a much larger crowd.

The sniffer in the doctor's or patient's hands already plays an indispensable role in evaluating how drugs perform in individual patients. Drugs that control insulin levels, blood clotting, or cholesterol are licensed only on the assumption that the patient's response will be carefully monitored, with prescriptions adjusted accordingly. The iPOP in your bathroom will be the next leap down this road. It will end up personalizing not just diagnosis but the molecular definition of both health and disease. And by doing that, it will personalize much of the evaluation of drug efficacy and safety. The clinical drug trial of the future will often be a personal trial—a trial of one.

ON ITS FLIGHTS to Haiti in the 1950s, Delta Airlines served exotic drinks adorned with a ten-inch swizzle stick. At the end of the stick, sporting a jaunty straw hat and hand-painted eyes, was a toy shrunken head. Which is pretty much where the sniffer takes medical diagnosis: to a biochemical dipstick doing the real work, with a doctor hovering nearby for decoration. The health care establishment has much reason to be dismayed. The sniffer threatens to do to their citadel what microprocessors and the Web will soon finish doing to much of the rest of the infosphere.

To begin with, sniffers give patients privacy—from nosy doctors. The introduction of home pregnancy testing kits in the 1970s, as one medical historian observes, reversed the "medicalization of pregnancy" and shifted the "locus of control at the moment of discovery." People may treasure privacy too much even to talk to a doctor, or they may reasonably lack faith in the what-happens-in-Vegas promises that doctors make about confidentiality. How about the government paymaster or private insurer that reviews the diagnosis and treatment before paying the doctor's bills? Or the nurse, file clerk, janitor, or taxi driver who finds the laptop that the doctor forgot?

Self-sniffing is cheaper, too—packaging diagnostic chemistry in a form suitable for sale from a rack located next to the gum and candy cuts the doctor and lab technician out of the loop. This diverts diagnostic dollars to Pharma, and with the money goes the doctor's ability to control demand for any treatment that a diagnosis might launch. And the sniffer gives the patient the information that the law too casually assumes he or she already gets. "Informed consent" for treatment may often be a hollow formality when a doctor who answers to a government paymaster controls all the informing.

The perfectly private sniffer will also instigate seditious thoughts and rouse the rabble. When the sniffer smells trouble, it will first send smart patients scurrying for a second, white-coated sniff. But as countless support-group websites already attest, many people are also eager to share whatever they can learn about their own problems with others in the same boat. And medicine as currently provided by the medical establishment begins to unravel from there.

Washington's medical science remains anchored in the study of crowds, because Washington's studies, licenses, and rules have to work wholesale, not retail. Thus in Washington, clusters of disease among the healthy expose contagious microbes and toxic chemicals, clusters of health among the sick establish that a drug is safe and effective, and clusters of a new disease among those taking the drug reveal its side effects. Until very recently, government statisticians were the best cluster spotters because they had the best access to the most data. The unlucky human bodies that constituted the early, telltale clusters served as sentinels and sniffers for the rest of us.

With their own sniffers in hand, however, the sentinels can now diagnose themselves and congregate spontaneously online. The perils of asbestos and HIV would have been recognized years earlier if people suffering from what had been two very rare cancers—mesothelioma and Kaposi's sarcoma—had been able to communicate as freely in the 1960s and 1970s as they do today. Google now spots incipient outbreaks of flu well ahead of Washington—a tool monitors for search terms such as "fever" and "chills" and ties them to the geographic area where the searches originated. Searcher anonymity is ensured, Google assures us, by the gnomes who run its search engines.

It's easy to dismiss Google medicine as lacking the discipline and rigor of real science, but as Stalin once remarked on the subject of tanks, quantity has a quality all its own. The individual patient is always the first to know what hurts, what hanky-panky he's been up to, what pills he's been popping, and whether they made him feel better or worse. Sniffers can elevate molecular self-diagnosis to Mayo Clinic levels of depth, detail, and precision. And as discussed in later chapters, patients, individual doctors, and medical Google gnomes already have, or will soon acquire, more power to gather far more data and make sense of it than Washington bureaucrats can ever hope to match.

Insurers have even more reason to worry. Private carriers survive only so long as their actuaries can predict average future costs for the crowd better than individuals can predict their personal prospects. Sniffers are all personal. Soon after the development of a test for HIV, the District of Columbia outlawed its use by life insurance companies, and the insurers immediately stopped selling policies to the city's residents, who now had the best information about how soon they were likely to die.

By transforming the ignorant, passive patient into an informed, meddlesome, and demanding customer, sniffers will also undermine every attempt to control costs by standardizing treatments. As discussed further in Chapter 13, government paymasters are now determined to use statistical analyses to script how doctors must prescribe drugs. Sniffers individualize the evidence and hand it directly to the patient. Every new sniffer that flags a new health risk, however remote, will spur demand for something to stave off the peril. As three doctors working with the VA Outcomes Group see it, America is already suffering from an "epidemic of diagnoses" and the "medicalization of everyday life."

The more patients learn, the more inclined they will be to take matters into their own hands, with the help of those doctors—Pharma's "whores," they've been called—who are willing to prescribe anything to anyone. There are many unhappy surprises in experiments of this kind, but in the boundless complexity of health and disease, good hunches and lucky guesses always play a large role, too. And in the search for luck, individual doctors and patients have an insurmountable edge over scientists and drug companies: they can perform far more experiments, they can break all of Washington's rules, and they can instantly share whatever they learn without waiting for clearance from the *New England Journal of Medicine*.

More important is that medicine's new power to track exactly what's going on deep down, whether it's wielded by patients or by their doctors, sets the stage for an entirely new approach to medical diagnosis and treatment, the antithesis of the Great Stink medicine that can't begin until a vile smell engulfs the city and people are dying in droves. Molecular medicine that begins with the power to sniff molecules leads to a mechanistic understanding of what those molecules do. It allows us to think about the rest of human chemistry the way surgeons began thinking about blood and oxygen centuries ago. The people in the blue scrubs don't yell for a statistician when a stretcher crashes through the doors of the emergency room, splattering blood on the ceiling—they just clamp and sew as fast as they can, because they know that blood-oxygen-brain means life.

Internal medicine took the first great leap down this path when Koch isolated V. *cholerae* three decades after Snow took the handle off the pump, and others then found tetracycline to kill it. Now we have biochemical tools to see problem proteins, lipids, and genes as easily as germs, and drugs are arriving, one by one, to tune and tame wayward chemistry long before it mushrooms into full-blown disease. And if doctors or patients can monitor problems down at the molecular level, they can track how well or badly the solution is working, and adjust treatment on the fly. The sniffer allows the audacious patient to find out quickly what works for the crowd of one. And when nothing works, the sniffer rallies the crowd of many to clamor for a cure.

The sniffer is the William Tyndale of our time. It translates the scripture of human life into text that ordinary people can at least read. And however difficult it may still be to decide what it all means, or what to do about it, reading it will certainly give ordinary people all sorts of medically heretical ideas . . . if the authorities allow it.

———

INVOKING A SNARL of federal and state laws, doctors, insurers, and the sniffing police are scrambling to limit patient access to the diagnostic sniffers that launch most of the rest of modern medicine. Whether these laws are good policy—whether they can even be squared with the Constitution—is a subject to which we will return in Chapter 18. The federal drug law does, however, empower the FDA to regulate diagnostic sniffers as strictly as it regulates drugs.

Walgreens was poised to start selling the Insight test kit when Stanford began iPOPing Snyder—until the FDA sent word that the kit lacked a license. The offending product would have allowed you to spit into a mail-in plastic vial in order to find out what your genes might have to say about all sorts of things, among them Alzheimer's, breast cancer, diabetes, blood disorders, kidney disease, heart attacks, high blood pressure, leukemia, lung cancer, multiple sclerosis, obesity, psoriasis, cystic fibrosis, Tay-Sachs, and going blind—and also how your body might respond to caffeine, cholesterol drugs, blood thinners, and other prescription drugs. Not long after, 23andMe, a company founded by the wife of one of Google's founders and funded by Google and Genentech, mixed up the genetic analyses of ninety-six clients and sent them inaccurate information about paternity, maternity, ancestry, and other matters. Invoking its authority to license every diagnostic "contrivance," "in vitro agent," or "other similar or related article," the FDA announced plans to crack down on all companies that attempt to sell such things to the public. State health officials have also piled on, by tagging gene scanning and other diagnostic services as the doctor-only "practice of medicine."

On the flip side, Washington indirectly mandates biochemical sniffing whenever it allows a drug license to be linked to a patient's biochemistry, or pays for a drug only when it's prescribed accordingly. Under the Affordable Care Act, Washington's dollar doctors will also require all insurers to cover in full the preventive diagnostic tests that it favors, and will put the economic squeeze on doctors and private insurers that choose to prescribe or cover others. Washington also recommends screening newborn infants for twenty-nine major disorders, and designates other tests as quite important. States make their own calls, based on the advice and vigorous advocacy of medical societies and patient groups. Depending on where he or she is delivered, your newborn will thus be subjected to a wide range of different sniffs that span congenital hypothyroidism, sickle-cell diseases, various metabolic disorders, and biomarkers that foreshadow problems with hearing or vision.

Few question the government's authority to police false claims about what a sniffer actually sniffs, and how accurately. The FDA is determined to make sure, however, that sniffers provide information that is not only "analytically" but also "clinically" accurate—to protect users not only from factually incorrect test reports but also from "unsupported clinical interpretations." Does Washington really have the right to proscribe sniffing because most of us just aren't wise enough to deal with biochemical facts until Washington tells us how, if at all, doctors and patients should respond to them? Can a sniffer dodge the FDA by supplying only a broadband torrent of accurate but uninterpreted data, leaving it to you to forward the file to Bangalore for a cheap read by doctors who don't answer to U.S. medical authorities? Several online gene scan companies are careful to insist that they sniff only for "educational" purposes, not to make a diagnosis. For now, the regulatory status quo betrays a chillingly authoritarian mind-set about the most important text you will ever have a chance to read.

3

INTELLIGENT DESIGN

IF YOU DOUBT that life sometimes owes its existence to intelligent design, chat with a woman who has survived invasive breast cancer, and with the biochemists who created Herceptin. The drug—a monoclonal antibody (mAb) designed by human genetic code embedded inside the cells of a mouse—targets a protein on the surface of breast cancer cells that conveys signals that propel the cancer's runaway self-replication.

In 1984, three scientists shared a Nobel Prize for developing the science that launched this radically new form of medicine. It is one of two revolutionary medical technologies developed in our generation that have limitless power to emulate and surpass life itself. Both mimic the human immune system—one inside live part-human cells, the other in the silicon of our computers. In the last three decades, biochemists have made the transition from blind guesswork to the systematic design of precisely targeted drugs. In doing so, they have transformed molecular medicine into an applied science anchored in immaterial code of almost limitless power and plummeting cost.

———

ON MAY 14, 1796, as on every other day, James Phipps' immune system was busy churning out millions of antibody proteins, each one bearing at its tip one of a billion possible variations of a biochemical sniffer, and dispatching them to prowl the boy's young body in search of something malodorous to attack. One bumped into a cowpox virus that Dr. Edward Jenner had lifted from lesions on the hand of milkmaid Sarah Nelmes. Responding to the alarm sounded by this tiny sentinel, other parts of the boy's immune system churned out armies of killer proteins and cells designed to destroy this same target—which, as it happens, is also found on the surface of the smallpox virus. In farmyards near his home, Jenner had stumbled across logic so fundamental that it has propelled pharmacology ever since. Hold a molecular blueprint of the disease up to a mirror, and you will see in the reflection the blueprint for a perfect antidote.

In a matter of days, a child's body learned how to mass-produce a drug perfectly matched to the most lethal disease on the planet. Until recently, your typical Monster Drug Company, powered by $150 billion of Wall Street's capital, could pull off a similar stunt once a decade, if that. The smallpox vaccine was the most valuable pharmaceutical product ever developed—its genius lay entirely in its ability to mobilize the process that nature had developed to design and mass-produce killer molecules.

Herceptin was designed and manufactured the same way, with a genetically engineered hybrid mouse-human standing in for the eight-year-old boy. And now, as the second revolution advances, computers are fast approaching the point where they can stand in for both boy and mouse. Tens of millions of patients with hypertension, HIV infections, and a wide variety of cancers are already being treated by drugs designed, in large part, by silicon.

In important ways, our synthetic immune systems can outperform the ones we were born with. The human body produces arrays of anti-molecules everywhere, not only to repel germs but also to detect flaws in our genes and sibling proteins and neutralize them before they become toxic or malignant. Most of the time these molecules are perfectly fitted to the rest of us—they don't poison our livers, shut down our kidneys, or trigger allergic reactions. But sooner or later they fail or turn against us. The ensuing problems are more difficult to deal with than most infections. Our own immune system usually refuses to attack chemistry it recognizes as kin.

Our synthetic immune systems, by contrast, will design bullets to hit anything we show them—the unusual chemistry of cancer cells, for example, or the chemistry of the human immune system itself. Some of the most successful mAbs are used to treat rheumatoid arthritis and other autoimmune diseases. And while nature took billions of years to design James Phipps, human ingenuity is now poised to overtake nature in a single generation. In the 1990s designers delivered a cocktail of new drugs quickly enough to overtake HIV, a virus that mutates so fast that most human immune systems can't keep up. Technologies now in hand have the power to design and start mass-producing a potent new anti-molecule in a matter of months. They will end up as swift as our immune systems, which complete the job in a few days.

Thus, our synthetic immune systems—of both the hybrid and the silicon varieties—are on a trajectory to become so fast and cheap that they can begin with a patient's own genes or proteins, or a close relative's, and produce a magic-bullet drug perfectly tailored to the patient's unique chemistry. We are moving swiftly toward systems that can design an anti-molecule to match any molecule found in the vast, complex, diverse, mutable library of biochemical code that defines humanity and all the rest of life on earth.

That's not the end of the story—far from it. But it's a good place to start.

––––––––––

IT IS IMPORTANT to understand how recently we acquired this power, and how fundamentally it has changed the drug design process.

For a century after Jenner, molecular-scale medicine made no further progress at all. Then Robert Koch and Louis Pasteur began tracking diseases back to infectious germs and, working independently, developed procedures for isolating, cultivating, and then killing or crippling every last germ harvested to create an effective vaccine safe enough to administer to healthy people.

In 1891, Paul Ehrlich joined Koch at his Berlin Institute of Infectious Diseases. Other biochemists had noticed that they could distinguish different types of bacteria under a microscope using a variety of colored dyes, each of which stained one group of microbes in preference to others. As Ehrlich mastered this work, he had the flash insight that may someday save your life. Different dyes had to be latching on to different chemical receptors on the surfaces of different microbes. The front end of a perfect drug would do exactly the same—it would be a bloodhound chemical that sniffs its way straight to a single target and disables or destroys it. This was "the dawn of nearly all modern pharmaceutical research," a medical journal would declare in a review of his work published in 2008, a century after Ehrlich was awarded the Nobel Prize for Medicine.

Ehrlich lacked the tools to isolate and examine molecular targets up close, but he was a good chemist. He began synthesizing and screening chemicals for antimicrobial activity, and came up with one that proved quite effective against syphilis. In their constant struggle for existence against each other, might germs themselves have designed magic bullets to fire at their rivals? Alexander Fleming's pursuit of that possibility, probably inspired by Ehrlich's ideas, culminated in the discovery of penicillin in 1928 and its commercialization two decades later. A team at Rutgers University then began to search for other such bullets secreted by soil-dwelling microbes and found streptomycin, another powerful antibiotic that cured tuberculosis.

Meanwhile, Gerhard Domagk, working at a German research institute established by a large chemical conglomerate that called itself the "Community of Interest of the Dye Industry," discovered that a red dye developed to treat leather protected mice and rabbits against lethal doses of staph and strep bacteria. The first patient successfully treated was a ten-month-old child gravely ill with what was, at the time, an invariably lethal staph infection. A few months later, before Domagk could conduct further tests in human patients, his six-year-old daughter Hildegard was accidentally infected with a virulent strep bacterium, and the infection spread to her arm and beyond. In desperation, Domagk, gave her a dose of dye, and she, too, recovered completely. The drug, however, turned the skin of both children a light red—Hildegard's permanently. Several years later, after the active piece of the dye's chemistry had been separated from the rest and more systematic studies had been completed, the story of Hildegard's astonishing recovery received worldwide publicity and helped launch sulfa drugs, the first major group of synthetic antibiotics.

When Hollywood made the movie in 1940, they called it *Dr. Ehrlich's Magic Bullet*. If you can see the structure of a target molecule, you can design a molecule to hit it. The designing can, in principle at least, be a perfectly systematic, logical process, from beginning to end.

IN THE CENTURY that spanned the discovery of the cholera bacterium in 1883 and HIV's arrival in Washington in 1981, vaccines and antibiotics did more to improve public health than all the rest of medicine combined, before or since. That makes it easy to suppose that the policies established to regulate them, which remain largely in place to this day, were probably good and worth keeping.

The short rejoinder, however, is that almost nobody was actually *designing* any drugs of any kind back then. The vaccines were derived from the disease itself, and they left the designing of the antidotes to our own immune systems. As an alternative to enlisting our bodies to design and mass-produce germicidal bullets, we then enlisted germs to supply most of the early antibiotics. Ehrlich's syphilis drug was mainly a product of luck, intuition, and guesswork; so were the sulfa drugs, and they didn't emerge until 1936, almost three decades later.

Insulin and estrogen, the two pioneering drugs that tinkered directly with human chemistry, had likewise been designed by nature first. Most of the small number of other useful people-tuning drugs that emerged before 1980 were developed mainly by hunch, trial, and error. Here, too, the alternative was to search for cures in nature's pharmacy. The first statin, which would lead to Lipitor and other cholesterol-controlling drugs, was isolated from a fungus in the 1970s by a Japanese biochemist, who figured that if some microbes secrete penicillin to kill rivals, others might have devised a way to kill by crippling production of the cholesterol glue that holds enemy cells together.

All of the vaccines and drugs of that era were also directed at quite simple progenitors of disease—simple enough to be isolated with the limited tools available at that time, and quite easily linked to medically important clinical effects. Bacteria, for example, are huge compared to viruses, usually mutate much less quickly, and can quite easily be linked to the infectious diseases that they cause. Insulin and estrogen were both isolated in the 1920s, and they, too, have certain direct effects that are easily tracked—on blood sugar (and hence on the clinical symptoms of diabetes) and on fertility.

The limits of the science thus ensured that medicine would remain focused on magic-bullet antidotes—most of them designed by nature itself—to diseases that were simple enough to succumb to magic-bullet assaults. The search for safe, effective antibiotics was further simplified by the fact that they were aimed at the chemistry of microbes, not people, and were therefore less likely to interact with the patient's body to produce unwanted side effects.

Even so, failure was the norm, not the exception. Until trials proved otherwise, the starting presumption had to be that the candidate drug would probably prove to be useless or downright harmful.

―――――――

OUR DRUG DESIGNERS, however, are no longer blind.

Enzymes are molecules that snap other molecules together or apart like Lego blocks. Drug designers guessed early on that they might be able to gum up enzymes by feeding them dummy blocks to snap away at. In the 1970s three researchers at Squibb set out to tame a protease enzyme that snaps proteins apart in the process of manufacturing a hormone that helps control our blood pressure. Their work culminated in the first major pharmaceutical product designed systematically from the very bottom up—Capoten, the first ACE inhibitor. The era of blind pharmacology ended officially on April 6, 1981, when the FDA licensed the drug.

Enzymes, like other proteins, are very long molecules that fold and snarl themselves into complex shapes. Just as microscopic pits encode a movie on a DVD, the structure's exposed surface encodes logic—in this case, the logic that allows each protein to perform one or more specific tasks on the assembly line of life. The code can be shut down by some sticky chemical that gums up the surface, much as a small glob of glop transferred from a kid's gummy finger makes the climactic scene on the DVD unplayable. But the chemical won't stick unless it is designed just right, to fit some key spot on the snarl.

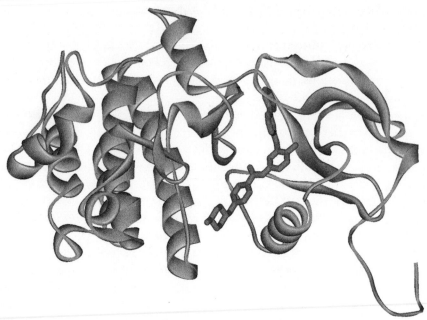

FIGURE 3.1   Gleevec bound to the BCR-ABL tyrosine kinase enzyme.
Source: Wikipedia, http://en.wikipedia.org/wiki/File:Bcr_abl_STI_1IEP.png.

The Squibb team knew of a snake venom protein that gums up the protease enzyme, and also knew that protease contains metal atoms that give it
a good grip on what it's about to snap. But nobody had yet worked out the
structure of the whole molecule. So the team designed a glob to fit a related
enzyme, which they figured was quite similar to the real target, and after
some further tweaking arrived at Capoten.* Emboldened by Squibb's success, other scientists would soon start designing another protease inhibitor
(saquinavir) that would help beat HIV.

Working out the sequence of the amino acids that constitute a protein
isn't difficult—the hard part is working out the structure of the snarled ball.
Today the first step is to persuade the balls to solidify into a perfect crystal
that's big enough to study. That's largely a process of trial and error, now
greatly accelerated by the use of computer-controlled lab robots, which
can try out many different protocols, largely unattended. Very high-energy
X-rays beamed through crystal provide a 2-D photograph of the 3-D struc-

---

*The structure of the real target has been worked out since, and it doesn't look much like
the stand-in at all; Squibb's team did launch a revolution, but it also got lucky.

ture; human brains assisted by powerful software then attempt to inflate the flat picture back into a 3-D model and work out how different molecules might latch on to the target's surface. The computers can't yet do it all on their own—the calculations are fantastically complex. But they are, in the end, just calculations, governed by deterministic laws of physics that control the atomic-scale chemistry.

Digital designers now also get help from hybrid silicon-human brains. Scientists struggled unsuccessfully for more than a decade to determine the structure of another protease enzyme involved in replicating HIV and other viruses. In 2011, researchers at the University of Washington turned to on-line gamers for help, by way of Foldit, a game that challenges players to fold proteins into the smallest biochemically possible space. Within three weeks the gamers had come up with models close enough to the real thing to allow the researchers to finish the job. The computers can't yet match the human brain's spatial visualization skills, but by way of networked game consoles they can engage and coordinate many nimble young brains in a process of working out the molecular structure of life.

Your teenager should enjoy, while it lasts, his right to ignore you because he's busy saving a million lives. When the computers that learned to beat the grand masters of chess surpass the callow wizards of *Halo,* as they inevitably will, they will be set to take charge of structure-based drug design from end to end. They will work out the structure of the target and then replace the brilliant biochemist's initial guesses about globs that might work with brute computational force, trying out millions of virtual gummy globs on a virtual ball of snarl. We know this approach can work. The virtual part aside, it's been done before, inside a youngster named Phipps.

IT'S ALSO ROUTINELY done now, inside the live cells that design and mass-produce drugs such as Herceptin. A mouse is vaccinated with a malign protein isolated from, say, the surface of a breast cancer cell. Cells in the animal's immune system undergo changes that allow them to churn out antibodies perfectly matched to that target protein, biochemists extract and cultivate the cells, and the antibody protein pours out, every molecule an exact copy of the same progenitor. The monoclonal antibodies thus produced can then be linked to biochemical teeth and claws of various types— radioactive components, for example, for use against cancer cells—or used directly to spur other parts of the patient's own immune system into action against the target.

The first mAb drug, which the FDA licensed in 1986, was an all-mouse antibody. Drug companies then began inserting human immune-system genes into the mice used to design mAbs. The "humanized" mAbs (such as Herceptin) that followed were hybrid mouse-human chemistry; then came hybrid monkey-human mAbs. "Fully human" mAbs are now in production; they are far better at dodging the patient's immune system and surviving long enough inside the patient to do their job. The mAb factory has been shrinking, too: in the 1970s it was a mouse; today, it may be a live culture of bacteria or genetically engineered human cells that have been immortalized by infecting them with a virus.

Nowadays mAb technology can also be manipulated to make random mistakes deliberately, just as our own immune system does, to create lots of variations in the sniffer chemistry that guides each antibody to its target. The human gene that codes for a key antibody protein, immunoglobulin, is repeatedly but imperfectly cloned; each slightly different copy is then inserted in viruses that use a familiar fungus, yeast, as a host. These collections of slightly different microbes then churn out large arrays of slightly different antibodies, from which those with desired properties are then selected.

Selling mAbs now earns drug companies tens of billions of dollars a year. Most target cancers and autoimmune diseases. Several hundred new mAbs are now undergoing clinical trials or awaiting FDA approval. Among their targets are a wide variety of cancers, cardiovascular and inflammatory diseases, multiple sclerosis, arthritis, asthma, and other degenerative diseases. The mAbs aim either to destroy hostile cells directly or to interrupt their ability to communicate with other cells. And as of 2009, more than twenty clinical trials were investigating mAbs aimed at microbes.

———

SO BIOCHEMISTS HAVE mastered two brilliant technologies for designing magic bullets. The rest of the story, in brief, is that our bodies and the diseases they spawn are very good at finding ways to dodge or disable them. So, too, are many infectious microbes.

Gleevec, for example, is widely viewed as one of the most magical of the modern bullets. Medicine now has "tools to probe the molecular anatomy of tumor cells in search of cancer-causing proteins," the National Cancer Institute exulted in 2001, when the drug was licensed to treat chronic myelogenous leukemia (CML). Gleevec is "proof that molecular targeting works." The orange pill has been called a "cancer antibiotic"—a daily dose, swallowed as easily as aspirin, returns white-blood-cell counts to normal ranges. But an important part of Gleevec's story concerns the ways in which it often fails.

In 1960, CML was the first cancer to be linked to a flaw in human genetic code, created when one of our chromosomes somehow trades its own short arm for the long arm of another to form the "Philadelphia chromosome." The gene in question codes for one of our many kinase enzymes, which, among other functions, help regulate cell division. The flawed version launches a frenzy of white-blood-cell production that kills the patient.

In his 2003 book *Magic Cancer Bullet*, Dr. Daniel Vasella, chairman of Novartis, recounts how scientists painstakingly designed Gleevec to suppress the rogue kinase. In even thinking to pursue this particular target, the designers displayed remarkable audacity. Nature often uses similar molecules to perform many different functions, in different cells and organs, as our bodies develop from a single cell in the womb through childhood and into adulthood. Kinase enzymes are found throughout our bodies, because one of their functions is to latch on to the body's key molecular fuel (ATP) and help it power the assembly lines.

Dr. Brian Drucker, at the Oregon Health and Science University, and a team of scientists at Novartis recognized that different kinases differ slightly in the structure of the binding pockets they use to snag ATP. They developed computer models to design various structures that might fit only the pocket in the kinase that causes CML, synthesized them, tested the most promising, and got to Gleevec.

But if subtle differences in kinases are all-important in designing Gleevec, we should not be too surprised to find differences in the chemistry that propels CML itself in different patients. And we do. Gleevec fails to help about one CML patient in ten. Or to put it another way, Gleevec never fails, it is the old medical taxonomy that has failed: CML is one name for at least two distinct diseases.

Then there's the problem that—like all cancers—CML keeps reconfiguring itself. About two out of every five patients on Gleevec benefit at first but then relapse because their cancer cells mutate into a Gleevec-resistant form. Drug designers have identified at least one of the mutations involved and redesigned other drugs to deal with it.

There's also a good-news flip side of both of these problems. In an appendix to his Gleevec book, Dr. Vasella includes brief, exhilarating accounts of how the drug, during early trials, stunned oncologists and late-stage cancer patients with its power to melt away tumors. The first happy account, however, is about Anita Scherzer, a patient brought to the brink of death not by CML but by an extremely rare cancer that develops in the stomach and intestines. Gleevec has since been widely used to treat ten cancers and other serious conditions that, to a clinician, look nothing like CML.

On reflection, this, too, should come as no great surprise. By reusing identical or similar building blocks to perform many different functions at different points in our bodies, nature sets the stage for drugs that cause nasty side effects. But we also find similar molecules, involved in what used to be viewed as entirely different diseases, and the same drug then cures them all. Gleevec, it turns out, inhibits the activity of three of the more than one hundred members of our kinase family. One propels CML, another gastrointestinal stromal tumors.

Biochemical logic likewise forces medicine to accept the brutal truth about uncommon side effects. They aren't flaws in the drug or unlucky accidents; they are uncommon variations in patient chemistry. The drug isn't defective, some of the patients are—or that, in any event, is how medicine now views the problem. It keeps the drug and votes the biochemically unsuitable patient off the island.

The broader lesson here is that every new, precisely targeted drug serves, in part, as a diagnostic instrument for exposing and classifying the molecular structure of diseases and the bodies that host them. As the drugs are developed and prove their worth, they redefine the diseases. In the medical literature, a steadily growing number of yesterday's diseases now come with prefixes or suffixes that designate some biochemical detail to distinguish different forms—"Ph+" (Philadelphia chromosome positive) in front of the "CML," for example. Similarly, drug labels and patient records increasingly include genetic and other biochemical details that determine the patient's ability to tolerate various drugs.

THE MAGIC BULLETS beat the easy problems first. The vaccines and antibiotics that wiped out the germs that killed us early allowed us to live long enough to be killed by our own, much more complex chemistry. Drugs that curb cholesterol and lower blood pressure postponed the heart attacks and strokes, giving our cells more time to mutate in ways that cause cancer. We now see rapidly rising rates of Alzheimer's, Parkinson's, and other nervous system diseases that often develop even later in life and appear to be even more complex. And more autoimmune diseases, which are particularly difficult because when we attack them we attack the system that protects us from so much else.

With these diseases, the gene and protein hunters now identify a surfeit of promising targets. Drug designers have the tools to design molecules that will subdue many of them, and lab tests often confirm that these targeted bullets are as magical as Ehrlich could ever have wished. But all too often, the drugs don't perform as hoped in clinical trials. It's the patients and their diseases that aren't cooperating. The magic bullets work one-on-one, but embedded inside live patients, the targets aren't druggable, or aren't worth drugging—at least not on their own.

We find out that a target is druggable, in the industry's circular jargon, when we manage to design a drug that can thread its way through the rest of a human body and hit the target without being destroyed by the body's own defenses or causing too much collateral damage along the way. That's the first place where the magic can fail. Our bodies are designed to repel and destroy intruders of any kind—including drugs. Our livers dismantle them, and the debris can be toxic. The drug or the debris sometimes poisons our kidneys. Our immune systems sometimes react so violently to intruders that they kill us, too.

We find out if a target is worth drugging by drugging it and watching . . . well, watching what? We can start tracking how things change down there—how the virus, or cancer cells, or plaques respond—almost immediately. But what we really want to know is whether the responses down there deliver the intended improvements in the patient's overall health. And here the magic can fail in all sorts of different ways. Late-stage cancers mutate so fast that they are rarely beaten by a single drug—cocktail cures are required instead. Or what looks like the same cancer to a clinician—CML, for example—can turn out to be two or more biochemically distinct diseases underneath, and different drugs will then be required to treat it in different patients. More generally, simple, static, one-size-fits-all design is, in the grand biological scheme of things, the path to extinction. Design life that way, and some other form of life will soon develop a magic bullet to turn it into food. Survival lies in complexity—the complexity of viruses such as HIV, which thrive by mutating very fast, or of human bodies, which include almost every kind of biochemical complexity imaginable.

As we shall see, the unexpected failures and successes of precisely targeted drugs now play a large role in exposing and unraveling the complex molecular dynamics of most of the diseases that we're still struggling to cure. Until quite recently, the biochemical reality that underlies the short epilogue to the tale of the Gleevec-CML magic was viewed as detail to be briefly noted on the Washington-approved label that accompanies every drug. The details are now taking over the show. We still speak of "developing a drug," but "developing the patients" would be more accurate. Both matter, of course—pharmacology isn't a science of one hand clapping—but all the complex biochemical details lie on the patients' side of the applause.

And that has far-reaching policy implications. The FDA certifies a drug's future safety and efficacy only insofar as the drug is used "under the conditions of use prescribed, recommended, or suggested in the labeling thereof." No drug gets licensed without a label, and the label is where the FDA, in effect, licenses future patients. That can't be done well until someone works out the relevant details of the patient-side chemistry. The FDA's current testing protocols weren't designed with that objective in mind at all. Nor was the elaborate structure we have in place to define the intellectual property rights that provide the incentives for developing the patient-side biochemical know-how. Nor were the rules that control how drug companies, doctors, and patients may collaborate to uncover and learn how to control the all-important details of the chemistry on the patient's side of the action.

4

COMING APART

ESTROGEN WAS THE first mass-market drug deliberately aimed at human chemistry. Beginning in the late 1930s, natural and then synthetic estrogens were widely used to treat hot flashes, insomnia, the discomforts of menopause, and a host of other conditions. In 1960, the FDA licensed Searle's Enovid—a combination of two precursor molecules that, when metabolized in the liver, yield estrogen and a second hormone that affects the uterus—as the first oral contraceptive. Hundreds of millions of young women have used estrogen-based pills since.

Pinning down estrogen's power to control fertility was easy, but estrogen has many other more subtle effects. Statistical studies began to link estrogen to strokes, heart attacks, and dementia when prescribed in conjunction with progestin (a synthetic form of the hormone progesterone); taken on its own, estrogen also increased the risk of uterine cancer. It appeared riskier when used by women who smoked or were overweight. On the other hand, when administered with progestin to postmenopausal women, estrogen apparently reduced the risk of osteoporosis, bone fractures, and colorectal cancer. "No other class of drug has been examined in such depth and for such a wide range of side effects, from blood clots and cancer to the production of ear wax," Carl Djerassi would observe fifty years after he led the development of the first oral contraceptive.

Biochemists have since identified many of the molecular receptors that allow estrogen to control biochemical processes that shape health for better or worse. To deal with the worse, we have developed *anti*-estrogen drugs. A daily dose of estrogen halts ovulation by tricking a woman's ovaries into believing she's permanently pregnant. Clomid, commonly prescribed to women who want to get pregnant, blocks those same receptors to persuade the ovaries to keep at it. So does tamoxifen, and it, too, is sometimes used as a fertility drug.

But tamoxifen was developed, and is more often used, to block estrogen receptors on cancer cells, because estrogen accelerates development of the estrogen-receptor-positive (ER+) forms of both breast and ovarian cancer. Estrogen itself, however, is used to treat other, estrogen-receptor-negative forms of cancer. Recent studies indicate that estrogen can be prescribed prophylactically to lower the incidence of breast cancer in some postmenopausal women. "The story of estrogen's role in breast cancer," an article in the *Journal of the National Cancer Institute* observed, "is starting to look like Dr. Jekyll and Mr. Hyde."

Defined by its clinical symptoms, breast cancer is a single disease that kills about forty thousand Americans a year. Oncologists, however, now categorize breast cancers by the presence or absence of receptors for estrogen, progesterone, and the HER2 protein. The first published reference to the "triple negative" form appeared in late 2005; it has since been the subject of hundreds of research papers.

Three yes/no receptor options make possible a total of eight receptor combinations. Multidrug therapies have to be tailored accordingly. They must often be changed on the fly, as the mutating cancer cells develop resistance to some drugs and susceptibility to others. The number of possible receptor combinations that affect how breast cancer might be treated will double every time another receptor is added to the mix by a new drug that can target it. Four yes/no receptor options would mean sixteen combinations, five would mean thirty-two, and so on.

To complicate the story further, our bodies often reconfigure drugs before they reach their targets. Over a decade after tamoxifen was approved, studies revealed that the drug itself wasn't much good at blocking estrogen—the effective blockers are produced when tamoxifen is metabolized by an enzyme in the liver. But significant numbers of women (with the numbers varying significantly across ethnic lines) have two copies of a gene that produces the enzyme in a form that can't activate the drug, and women who have one copy activate much less of it.

Most of the molecular details about how estrogen affects different bodies emerged only in the last decade. Much of the research got under way because of the discovery that different patients responded in such different ways to the same drugs. All the details hinge on the different receptors and enzymes found in different cancers and livers in different bodies. So much for the view that doctors define a disease and biochemists then develop a magic-bullet drug to beat it.

When dealing with breast cancer, oncologists now speak of treatment "algorithms": sets of rules for selecting among a broad array of drugs combined in a broader array of cocktails. A consensus statement released by breast cancer specialists in 2009 conceded that "the previous attempt to produce a single-risk categorization and a separate therapy recommendation are no longer considered appropriate." In its place, doctors should craft personalized therapy, guided by the biochemistry and the several categories of drugs now available. "We expect the refined algorithm to change clinical practice because it clarifies the indications for each treatment modality available today." Three years later, a new study linked the disease to genetic profiles, several of them new, that define "10 quite distinct diseases." Biochemists and oncologists now have in hand "a new 'molecular map'" to guide both treatment and the development of new drugs.

––––––––––

MEDICINE'S ABILITY TO develop these molecular maps fundamentally changes how we should view medicine's—and Washington's—venerable focus on clinical symptoms and its reliance on purely statistical correlations to link causes and effects.

In 1713, the Italian physician Bernardino Ramazzini noticed that Padua's nuns were more prone to breast cancer and less prone to cervical cancer than were other women. He correctly attributed both differences to chastity. Most cervical cancers, we now know, are caused by the sexual transmission of human papillomavirus (HPV). But sex also causes pregnancy, which triggers various hormonal changes that lower a woman's long-term risk of breast cancer.

Contraception and abortion prevent pregnancy almost as well as chastity does, and their availability can also influence behavior in ways that increase the risk of exposure to HPV. Thus, if framed just right, a statistical study can reveal links between two cancers and the choices women may make to prevent or terminate pregnancy. Many experts insist there are no such links, and they, too, have statistics to prove it. Some concede that there's no link "after adjusting for known risk factors." But one of those factors is postponed pregnancy. One might equally well say that, "after adjusting for exposure to HPV," unprotected sex doesn't raise the risk of cervical cancer. That statement is in fact true, but it is outrageously misleading, in that most HPV exposure is caused by unprotected sex.

Sophistry of this kind illustrates how statistics can be misinterpreted or manipulated when they deal with a cause that is anything but the final link in a long chain of causes. Ramazzini couldn't see hormones or viruses, so he linked two cancers to the chastity that he *could* see. Today, with the underlying details well understood, the statistics of remote potential causes serve only to conceal and mislead. Contraception or abortion can indeed be one remote link in a complex causal chain that ends with breast or cervical cancer. But here again, it borders on scientific fraud to leave things there when we can discern the intervening details. If the purpose is to help women make healthier choices, they must be informed of the causal details that are involved.

As they did in 1713, statistical correlations still play a major role in exposing the causes of diseases and assessing the performance of drugs. But now such correlations launch the search for mechanistic biochemical explanations. The germ's, gene's, or drug's ability to cause a precisely defined effect in a precisely defined biochemical environment is a fact that science can certify with certainty. Every fact of this kind narrows the realm where statistical studies of the clinical symptoms of crowds still rule. Every advance in molecular medicine moves us another step away from the medicine of the mob.

By analyzing patient biochemistry, prescribing precisely targeted drugs, and tracking how patients respond, medicine tracks diseases back to their origins, and disassembles them as it does. The genesis of breast cancer has been linked to at least four different genes. Women with certain variations in a gene that normally plays a role in repairing genetic errors are likely to develop cancer in at least one breast. In 2011, a group of researchers published the results of a complex "Bayesian network" analysis (more on this type of analysis in Chapter 11) that points to seven more common genes that appear to play key roles in triggering the disease, and numerous others involved in driving the tumor's growth.

The same process takes apart the rest of the patient, too. Tamoxifen works right if you have the right genes to metabolize it. Tolerance for many other drugs seems to depend on a couple of thousand variations in a couple of hundred genes. Eleven variations in just one gene affect responses to common antidepressants. White Americans have less tolerance for some antidepressants, antipsychotics, and heart disease drugs than do members of other demographic groups, while blacks respond poorly to certain drugs for high blood pressure.

Molecular medicine is also able to determine when to leave well enough alone. A gene variant discovered in early 2008 apparently protects about 40 percent of African Americans as well as certain heart-disease drugs do, by tinkering in much the same way with adrenaline's effect on heart cells. Middle-aged men with long telomeres—repetitive sequences of DNA that protect the ends of chromosomes and choreograph the eventual death of our cells—develop heart disease about half as often as short-telomere men, and may therefore be better off skipping certain prophylactic drugs. Adriamycin had been routinely used to treat late-stage breast cancers after surgery, but it can cause serious heart problems and spawn other cancers. A test that profiles twenty-one genes now identifies patients who will do as well on milder drugs, and others who can skip chemo altogether. A fourteen-gene test of

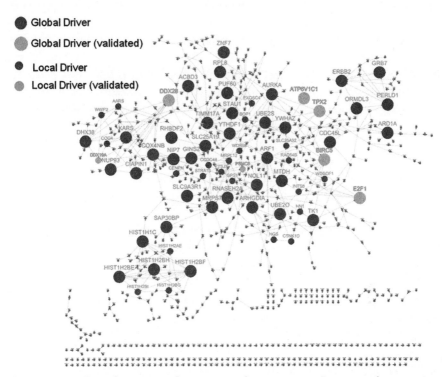

FIGURE 4.1 **Causal genomic alterations in breast cancer. Larger nodes represent global drivers of tumor growth, medium-sized nodes represent local drivers, and the smallest are nondrivers. The darker nodes represent cancer susceptibility genes.**
Source: Linh M. Tran et al., "Inferring Causal Genomic Alterations in Breast Cancer Using Gene Expression Data," *BMC Systems Biology* 5, no. 121 (August 1, 2011).

lung cancer cells can accurately identify the patients who need chemotherapy after surgery. A four-gene signature enables doctors to identify with high accuracy the prostate tumors that require aggressive treatment, and may thus allow many men to skip brutal treatment they don't need.

And that's it. Two thousand years after Democritus postulated a material universe made up of indivisible atoms, science can now track life down to its basic elements. Health once depended on four fickle humors, which apothecaries rebalanced with eye of newt, adder's fork, and fillet of the fenny snake. Cholera was just one ill-humored disease among many that left patients lying in puddles of their own waste—then Pasteur found V. *cholerae* in the puddle, and others found tetracycline to kill it. Until recently, Adriamycin's toxic side effects were a regrettable but necessary risk that too many patients had to take—then scientists invented gene sniffers to weed out the patients who should skip the drug.

At each step, medicine has advanced by disassembling the old swellings, fluxes, and fevers into hundreds of distinct germs, the new diseases into thousands of genes, proteins, and other biochemicals. Pharmacology has refined its power to target smaller shards of hostile life while dodging friendly molecular bystanders. Our biochemists will continue to search for such shards and develop drugs to control them for as long as life keeps inventing them.

It will be a long search. The still common, politically comfortable assertion that in our genes we are all "99.9 percent alike" is based on the science of a decade ago—ancient history, in this field. Completed in 2003, the first decoding of "the human genome" was in fact a decoding of about two-thirds of the genome of a single man of European descent plus a hash of a handful of other genomes. When statistical analyses kept pointing to hereditary strengths and weaknesses that couldn't be tied to known genetic differences, gene hunters began searching for more differences and were startled by how many they found.

Researchers investigating the wild mutability of cancer cells recently discovered that humans share with apes a biochemical quirk that introduces "copy number variations" (CNVs) when our cells replicate. They also show up during the creation of our reproductive cells, and then show up in our children. CNVs remove parts of genes, entire genes, or multiple genes, in stretches of chromosomes long enough to be seen under a microscope. About one thousand genes are duplicated in most people, about half of those in numbers that vary from person to person. The discovery of CNVs, in the words of one geneticist, has lifted the veil "on a whole new level of genetic diversity."

The large genomic gaps created by CNVs are now being linked to an elevated incidence of a growing number of disorders—among them nervous system disorders such as autism and schizophrenia and autoimmune disorders such as lupus—and have direct implications for how these might be treated. With genes, as with drugs, dosages matter. Almost all of us, for example, carry multiple copies of certain immune system genes—but just how many may affect how susceptible we are to HIV infection or Crohn's disease, or help determine how likely we are to gain weight. A mutation in a hemoglobin gene protects against malaria, but a child who inherits a copy from each parent will develop sickle cell anemia. The CNV diversity engine also protects cancerous tumors from our drugs—as mutations multiply, the magic bullet that wiped out a cancer cell's ancestors stops working and the disease gets progressively harder to beat.

Each of us also carries thousands of genetic spelling errors—single nucleotide polymorphisms (SNPs). In October 2012, the 1000 Genomes Project reported that its study of fourteen population groups in Europe, Africa,

East Asia, and the Americas had identified thirty-eight million SNPs, with more to come shortly. Scientists had already used information from the study to identify one SNP that raises the risk of multiple sclerosis and explains why the disease may be aggravated by certain drugs that are used to treat autoimmune disorders.

Another study, completed a few months earlier, analyzed SNPs in the potential "drug target genes" of fourteen thousand individuals thought to be particularly susceptible to heart attacks, strokes, obesity, and other health problems. On average, each subject was found to carry about fourteen thousand SNPs, about twelve thousand of which were exceedingly rare. Each subject carried an estimated three hundred genes with variants found in less than 0.5 percent of the population that would probably disrupt a protein's structure in ways likely to undermine health and affect how the protein would respond to drugs. Most of these rare variants, as the *Science News* report on the study put it, are "practically secret family recipes. Others reveal the distinct flavor of geographic regions, much like wines or cheeses."

Because rare variations tend to cluster geographically, even large statistical studies of a disease will often miss many of them, the authors of the study conclude: "Surveys of common variants are only observing a small fraction of the genetic diversity in any gene." And the common variants usually have only weak links to the disease—they wouldn't have persisted long enough to become common if the links were strong. So the best that drug designers can do is search for disease links to what have been called "Goldilocks genes"—variants that are common enough to show up in statistical studies and also have a fairly strong causal connection to the disease. The drug target study of variants related to heart disease and diabetes found 105 Goldilocks genes in Europeans, 132 in South Asians, and 210 in African Americans.

But the Goldilocks genes may well account for a relatively small fraction of the total genetic risk. Seemingly common disorders, it now appears, often reflect large numbers of rare, distinct flaws that cause the same clinical symptoms. They may be different flaws in the same protein, or flaws in two or more different proteins that must work together to maintain health. Hundreds of different proteins that control the interfaces between nerve cells, for example, can apparently play a role in choreographing Alzheimer's, Parkinson's, epilepsy, and more than 130 other brain disorders. Autism has been linked to a few common genetic variations, but also to a much larger number of rare ones, which researchers now believe play dominant roles.

Our molecular diversity has, finally, one more important dimension that biochemists have only just begun to unravel. Until recently, most of our DNA had no known function and was often referred to as "junk." Now we know that a large part of it codes for lincRNAs, a group of perhaps five thousand to ten thousand nucleic acids that carry or suppress important messages, control where proteins go, and play other essential roles in putting the pieces together and coordinating how they interact. Certain uniquely human lincRNAs, for example, are activated only in our brains. Another lincRNA proliferates aggressively in breast cancer cells and appears to play a role in deciding when they drift away from the primary tumor.*

———

No one knows why we humans and our simian ancestors came to incorporate such a strong engine for genetic differentiation, but our tendency to move on and then settle down in a broad range of different environments has certainly had something to do with preserving some of the variations the engine spawned. The adaptations that proved advantageous to our ancestors, however, often cause problems for us in the very different habitats that we occupy today.

Europeans and Asians, particularly in China and Papua New Guinea, carry HLA immune-system genes acquired tens of thousands of years ago by interbreeding with the now extinct Neandertals and Denisovans. These genes, which are not found in modern African populations, probably spread and persisted because they allowed some of our distant ancestors to survive microbes found only in northern latitudes. But HLA genes have also been linked to the onset of type 1 diabetes, celiac disease, and other autoimmune diseases, as well as autism and other disorders. Screening donors and recipients to ensure matches of these genes is also essential when transplanting organs and other tissues.

When our immune systems couldn't do the job, other genetic variations emerged to create biochemical environments hostile to infectious interlopers. Your ancestors survived bubonic plague in medieval Europe? Cholera in Victorian London? Malaria in South Asia? Sleeping sickness in central Africa? You are more likely to carry genes that improved your ancestors' ability to fend off each of these diseases, but that (in the same order) can cause a lethal disease associated with iron surpluses in many parts of the body; or increase your risk of developing heart disease; or cause you to develop sickle cell anemia, or to retain unhealthy amounts of salt and iron that increase your risk of heart disease, stroke, or liver disease; or elevate the risk of chronic kidney disease.

———

*Future references to "genetic" factors in this book should be understood to include lincRNAs and other "epigenetic" factors that control gene expression, along with the many variations in protein and other chemistry that they entail.

Our ancestors also adapted in different ways to different social and environmental factors. For most of its history, agriculture was bad for human health—people grew smaller and died younger as their diets lost variety. In small but significant ways, genetic profiles changed. Pacific Islanders and other communities that stuck with hunting and fishing were much healthier but now deal less well with modern diets. Communities living in different geographic areas also adapted to the different toxins that plants have developed to repel microbes, insects, and herbivores. People who still carry the favored genes—genes quite similar to those that allow certain insects to thrive on toxic plants—are apparently better able to deal with nature's poisons and certain drugs. And climate played a role. White skins are far more prone to develop malignant cancers because they lack melanin, which intercepts DNA-shattering sunlight; people with darker skins that block sunlight are more likely to have vitamin D deficiencies associated with a higher risk of heart disease. Low insulin levels and high blood sugar can significantly improve our ability to survive extreme cold; if your ancestors lived in cold climates, you are more prone to develop adult-onset diabetes.

Other genetic frailties that now cause us much harm reflect a different aspect of modernity: we live too long. Until recently, there was little evolutionary pressure to weed out grave genetic errors that occur later in life, because most of us died much earlier. So our species came to rely on biochemical processes that are useful for a time but so prone to error that they will probably kill us if we live long enough.

Our immune systems rely on constant, rapid, and random differentiation throughout life to keep pace with the constantly mutating microbial world. This process can go wrong in limitless ways, and when it does it may tag some part of our own chemistry as alien, and we develop multiple sclerosis, lupus, rheumatoid arthritis, or some other autoimmune disease. Many of our cancers seem to be triggered by the accidental resurrection of dynamic and constantly changing biochemical processes that spawn brains, bone marrow, and all the rest of the complex, resilient biochemistry that turns a single cell into a baby and a baby into an adult, and then keeps us healthy for most of our lives. Young children, unlike their parents, can regrow their fingertips. Most cancers appear to be launched by mutant stem cells, which behave much like embryonic cells in the womb. They keep reconfiguring their own chemistry. They sustain themselves by seizing control of healthy tissues wherever they land. And when they drift from their place of birth and land elsewhere in the body, they start the process all over again. Pregnant mothers occasionally transmit their malignant skin cancer or leukemia to their unborn babies.

More generally, the health improvements that have led to the rapid expansion of global human population over the course of the last two centuries have allowed harmful genetic variations to proliferate faster than they are removed by selection pressures that, over the long term, eliminate genes that do more harm than good. Viewed from an evolutionary perspective, genetic flaws that seriously undermine health are likely to be rare, of recent origin, and clustered in families and small geographic communities. And that is exactly what the gene hunters are now finding.

ALL THIS GENETIC diversity creates a human biosphere that is far more difficult to fathom than statisticians once thought. Until recently, geneticists believed they could predict the average level of gene-to-gene human diversity by examining about ten thousand individuals. The high frequency of extremely rare, geographically clustered variants now indicates that they would have to examine far more people than that.

Sophisticated analyses of sufficiently large amounts of data can, nevertheless, smoke out how a large number of molecular factors can combine in many different ways in different people to produce what, when viewed from the top, looks like a single disorder. A predictive engine developed from a high-powered statistical analysis of 1,313 SNP variants, for example, can predict the lifetime risk of the most common form of stroke in white patients with better than 80 percent accuracy. Another model provides even higher accuracy in predicting the stroke risk in patients with sickle cell anemia, most of whom are black. The core of the engine is, again, a Bayesian network that maps out links and quantifies interactions among SNPs. Figure 4.2 shows only the innermost circle of the network, which includes thirty-seven SNPs in twenty genes that affect the risk of stroke directly. Analyses such as this one then point the way to a range of prophylactic multidrug regimens matched to the different combinations of risk factors present in different patients.

Simple statistical studies that focus on a large group of patients who share a common set of clinical symptoms will, by contrast, often conclude that there's no single target down there worth hitting. All such studies and conclusions reflect the old magic-bullet model of a single simple cause directly linked to simple effect. A high level of complex and statistically elusive biochemical diversity likewise makes it impossible to assess a drug's efficacy or most of its potential side effects using a conventional, one-dimensional statistical analysis of data acquired in biochemically indiscriminate trials.

To judge from what our drugs are revealing when tested this way, most of the diseases we're struggling with today are far more biochemically complex than the diseases that medicine has mastered so far. If a single drug works

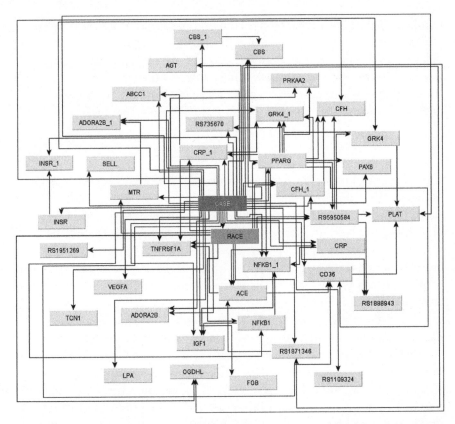

FIGURE 4.2 **Bayesian network describing the joint association for thirty-seven single-letter genetic variations and race with the risk of cardioembolic stroke.**
Source: Rachel Badovinac Ramoni et al., "Predictive Genomics of Cardioembolic Stroke," *Stroke* 40, supp. 3 (March 2009): S67–S70.

at all, it works only in a minority of patients. The development of a first, somewhat effective drug that manages to squeak through the licensing process often plays an essential role in exposing the fact that the disease comes in many biochemically distinct forms. The development of additional drugs disassembles the disease still further. This process then often leads not to a magic-bullet cure but to arrays of different drugs and algorithms for deciding how to prescribe them in different ways to different patients . . . if we're lucky.

Many of today's seemingly intractable disorders may have no single point of vulnerability—in which case no single drug will ever help much at all. Late-stage cancers often dodge all our drugs by reconfiguring their own chemistry on the fly; they can be significantly slowed only by drug cocktails that attack them at multiple different points. Cancers and many other complex disorders thrive by hijacking the redundant systems deployed throughout our bodies to protect us from attacks on mission-critical molecules. Here, too, effective treatment may require a coordinated attack with multiple drugs. All the details—which drugs to use and what side effects to worry about—may, again, depend on many different variations in the patient's biochemistry.

———————

EHRLICH WAS RIGHT: drugs must mirror life. But life is a vast tool kit of molecules that mutate much more often than we once supposed, are constantly being remixed in new combinations, and operate in complex chains and synergistic webs of biochemical reactions. To beat biochemically complex diseases we will need a pharmacy stocked with a correspondingly large and diverse array of drugs, together with complex protocols for prescribing complex treatments to match the many different ways in which drugs perform in different bodies.

In our war against germs, we will stay ahead of life only by developing new drugs faster than the germs can develop bullet-dodging genes. In our war against our own rogue chemistry, there aren't going to be many more drugs like Capoten and Lipitor, drugs that successfully target a widely shared problem and perform in much the same, safe way in a large majority of human bodies. We will run out of blockbuster, one-size-fits-all, single-drug antidotes to problematic human chemistry, just as we ran out of one-size-fits-all legacy germs to conquer, because nature shuns one-size-fits-all life.

As it disassembles our bodies, molecular medicine is, indeed, erasing all tidy lines between health and disease. Those lines seemed simple and clear during our long war against the legacy germs, but they disappear when medicine targets retroviruses, cancer cells, prions, genes, and proteins that are spawned by or hijack our own chemistry and turn it against us. In light of what we now know and are now capable of doing with drugs, we can no longer speak of "normal health," or link any wise or useful public policy to that notion. A newborn baby in the pink of health isn't pink enough to ward off measles or polio. Many bodies aren't healthy enough to resist the seeds of cancers, autoimmune diseases, and countless others implanted inside them at conception. And none can ward off the omnibus disease we call aging; any systematic assault on that disorder will likely require a broad array of medicines prescribed in ways that mirror all the ways in which our differentiated cells deteriorate.

Beating fast-mutating germs, breast cancer, the processes that clog our brains with plaques, or our own immune cells when they turn against us gets increasingly difficult as the diseases embed themselves deep inside cells, organs, and tissues and spread across the body. But if we go after them early enough, there is no scientific reason to suppose these disorders are unbeatable. All the science is headed toward antidotes that work as Jenner's did, anticipating and intercepting problems before we even know we're sick. Nature has already shown that it is biochemically possible to ward off many diseases that medicine still views as hopelessly difficult. Certain combinations of genes enable some people to control HIV infections on their own. Other genes confer a high degree of general immunity against cancer—so something down there must be good at mopping up genetic errors before they spiral out of control. Many brains don't develop Alzheimer's. Studies of the exceptionally old have identified a number of "Methuselah genes."

If nature can do it, there's a good chance that we can find a way to do it, too. Drug designers are now taking their magic-bullet science to the next level, reassembling the pieces to create treatments that mimic the biology of good health in all its complex molecular diversity. They have already had some stunning successes that hold great promise for the future.

REASSEMBLING THE PIECES: PART 1

WHEN THE GENETIC facts finally surfaced, biologists were surprised to learn that a deer is more closely related to a whale than to a pig. Elephant shrews, golden moles, manatees, and elephants likewise have much in common, deep down. So do humans, apes, rabbits, and bats. Today's taxonomists no longer infer kinship from the large-scale anatomical and behavioral features that guided their predecessors. As geneticist Steve Jones puts it in *Darwin's Ghost,* "The nineteenth century used bones, but now we have molecules."

Medicine's descent into the microcosm is creating completely new molecular taxonomies of humanity itself, in its sickness and its health. And it is by combining the molecular science of human bodies with a molecular understanding of how drugs work that we are arriving, for the first time in history, at truly scientific medicine.

Or at least we are getting as close to the truly scientific as medicine can ever get. Viewed as a whole, every patient is biochemically unique, while science addresses regularities and repeatable events. But starting with one genome assembled at conception, nature crafts our diversity using and reusing many of the same or similar molecules throughout each body, and many similar combinations of molecules show up again and again in different bodies. By pooling molecular and clinical data, medicine can identify those combinations and work out how they function to shape our health.

This is where molecular science reassembles the pieces that the previous chapter described. The integration of drug and patient molecular science is now giving pharmacology a predictive power far greater than the molecule-blind crowd science of the past could ever deliver.

———

Consider first how the development of a novel drug advances the rest of medicine and benefits competing drug designers. The new drug works as well as it does because it hits a new, well-chosen target—a receptor on a certain type of breast cancer cell, for example. When clinical trials confirm that the drug has few nasty side effects, they establish that the receptor must be quite specific to the disease, and can be targeted without risk of sideswiping bystanders elsewhere in the body. The pioneer's success thus reveals important biochemical facts about how human bodies work.

Copycat drugs often arrive in the market soon after: competitors modify the chemistry just enough to dodge the original patent, often improving it when they do. A Japanese researcher, for example, isolated the first statin drug, compactin, from microbes, and a colleague established its efficacy in limited human trials. Following that lead, others quickly synthesized and commercialized slightly different versions of the molecule that had fewer side effects and worked better.

As we have seen, the selective efficacy of precisely targeted drugs then helps medicine disassemble the targeted disease into a cluster of biochem-

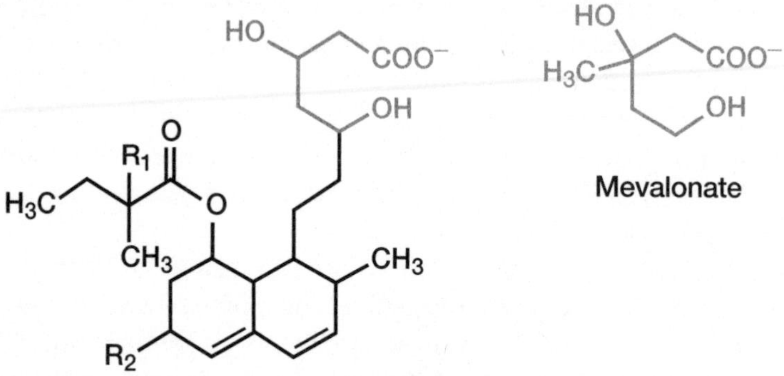

$R_1 = H$	$R_2 = H$	Compactin
$R_1 = H$	$R_2 = CH_3$	Lovastatin
$R_1 = CH_3$	$R_2 = CH_3$	Simvastatin
$R_1 = H$	$R_2 = OH$	Pravastatin

FIGURE 5.1   Statin drugs: original and copycats. The drugs suppress cholesterol production by binding with an enzyme that converts mevalonate into sterols, a key stepping-stone in the production of cholesterol. Akira Endo isolated compactin from a fungus; Merck changed it slightly to create the first licensed statin drug, lovastatin, and then again in a slightly different way to create simvastatin. Bristol-Myers Squibb followed with pravastatin.

Source: Akira Endo, "A Gift from Nature: The Birth of the Statins," Nature Medicine 14 (2008): 1050–1052.

ically distinct diseases, driving the search for more drugs to fill in the gaps. And new drugs simultaneously help expose molecular features shared by diseases that look quite different at the clinical level. An oncologist with a hunch about the underlying biochemistry and a patient willing to take the risk, for example, try using Gleevec on gastrointestinal cancer and find that it works stunningly well. Time and again, the benefits of developing a new drug far surpass original expectations when it turns out that the drug's target is involved in propelling several different diseases.

These valuable spillovers also occur as medicine maps out the biochemistry of side effects. The discovery that a certain liver enzyme, which most patients have but some lack, converts tamoxifen into the components that do the real work gives medicine one more genetic biomarker to work with. Nature didn't design us with tamoxifen specifically in mind, so the enzyme probably has some more general ability to chop up or reconfigure molecular interlopers. The genetic test that now helps doctors prescribe tamoxifen more accurately will thus undoubtedly help guide the prescription of other drugs, too.

The same process works in reverse—the more we learn about the detailed biochemical structure of human bodies, the easier it gets to design drugs to fit specific biochemical profiles and prescribe them accordingly. To generalize Ehrlich's great insight, disease science advances drug science and vice versa. Because each mirrors the other, they evolve in tandem. The better we coordinate that process, the faster we develop what we really need—an integrated drug-patient science.

———————

BY POOLING MOLECULAR information about existing drugs, their targets, and their intended and unintended effects, researchers using powerful mathematical tools are already learning a great deal about the structure of the underlying biochemical networks, and how those networks are likely to affect the performance of future drugs.

To the eyes of a biochemist, the measure of a drug's beauty is how likely it is to hook up smoothly with its target. In a 2012 paper, "Quantifying the Chemical Beauty of Drugs," one research team described how it pooled information about multiple aspects of the molecular structures of drugs successfully licensed in the past to arrive at a general algorithm for predicting the likelihood that a candidate drug will be successfully absorbed by the human body and won't have toxic side effects. The team used similar tools to quantify the beauty of potential binding sites that a new drug might attempt to target.

Another research group combined a catalog of 809 drugs and the 852 side effects known in 2005 with information about each drug's chemical

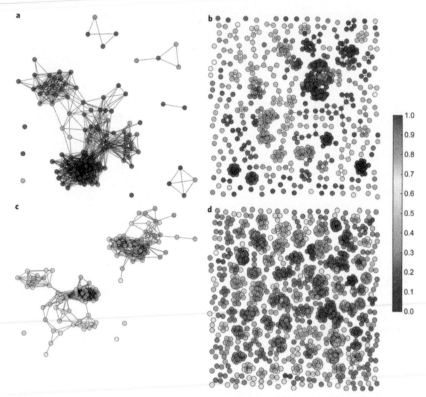

FIGURE 5.2 Charting the "drug-likeness" of different compounds. Each dot represents a drug known to interact with one of four different targets. Tight, connected clusters indicate chemical similarity, with the darkest dots representing the most promising druglike compounds. Biochemical properties determine whether a drug is likely to reach its intended target (bioavailability) or cause unwanted side effects. By comparing the molecular structures of different drugs, drug designers can use clinical experience with the old drugs to predict how the new drug will perform.
Source: G. Richard Bickerton et al., "Quantifying the Chemical Beauty of Drugs," Nature Chemistry 4 (January 24, 2012): 90–98.

properties and targets. Network analysis software was then able to predict almost half of the additional side effects that have emerged since then. "We were pleasantly surprised," said Ben Reis, director of the predictive medicine group at Children's Hospital Boston. Part of the network's power comes from the inclusion of information not previously considered in attempts to assess side-effect risks—the drug's molecular weight and melting point, for example, and what specific part of the body the drug targets. As Reis notes, "The network encodes a lot of information from other worlds."

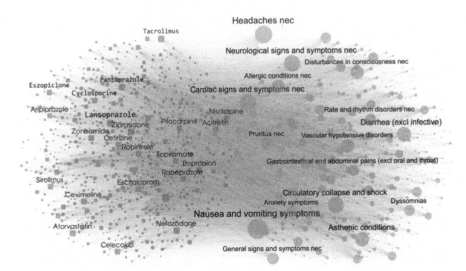

FIGURE 5.3   **Predicting adverse drug events using pharmacological network models.**
Clinical experience links drugs to known side effects (on the right); biochemical
properties link drugs to other drugs (on the left). The network then reveals new
links between drugs and side effects.
Source: A. Cami et al., "Predicting Adverse Drug Events Using Pharmacological Network Models," *Science Translational Medicine* 3 (December 21, 2011), doi:10.1126
/scitranslmed.3002774.

The team is now investigating what types of biochemical data have the
most predictive value, and studying drug-drug interactions. "We're moving
from a paradigm of detection—where it takes sick people to know something is wrong—to prediction."

By mining ten years' worth of clinicians' notes on the treatment of
forty-seven hundred patients at a large psychiatric hospital, another team
uncovered some eight hundred unexpected pairings of health problems.
Adding gene and protein data relevant to about one hundred of these pairs
revealed previously unknown molecular connections between such things
as migraines, hair loss, gluten allergy, and schizophrenia. Yet another team
developed what one member describes as an opposites-attract dating service for drugs and diseases. Using public databases that contain thousands
of genomic studies, the digital matchmaker searches for diseases that push
a specific human biochemical north, and drugs that push it south. Early
results suggest that a drug currently used for epilepsy might also be useful
in treating certain inflammatory bowel disorders, while an ulcer drug might
also help treat some forms of lung cancer.

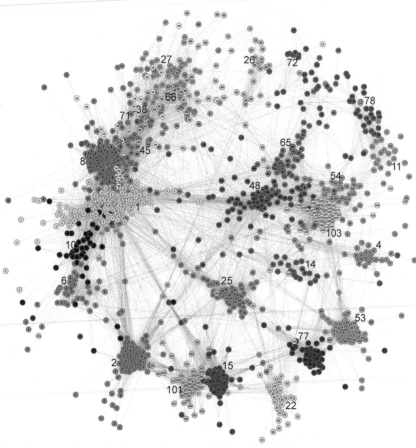

FIGURE 5.4 Using electronic patient records to discover disease correlations. Overlapping or adjacent conditions such as diabetes (number 26 at top) and hypertension (number 72) suggest a likely molecular kinship. Comparing patient biochemical profiles can then reveal the specific biochemical processes involved.

Source: Francisco S. Roque et al., "Using Electronic Patient Records to Discover Disease Correlations and Stratify Patient Cohorts," PLoS Computational Biology 7, no. 8 (2011): e1002141.

Biochemists call this "repurposing." Gleevec is used to treat both leukemia and gastrointestinal cancer. Rogaine and Viagra weren't developed for new hair and better sex; only in clinical trials did patients unexpectedly report those benefits. AZT, designed to treat cancer, was used to fight HIV. Gemzar, developed to counter viral infections, is used to treat cancer. Repurposing is as close as pharmacology gets to a free medical lunch. Nobody has to reinvent the drug or the factory that mass-produces it. Much is already known about the drug's safety and side effects. And many more cut-rate lunches are out there waiting for us. Life has been repurposing molecules from the beginning, so we now find identical or similar ones scattered all over the place. And when we find them, we find new commonalities among diseases that point to new uses for old drugs.

———

AT THE FOREFRONT of the integrated drug-patient science are the germs that we host. Each germ presents a discrete and biochemically simple target that can be cultivated, analyzed, and genetically manipulated in the lab to reveal much about how molecules give a microbe its power to infect us, thrive inside us, and succumb to—or shrug off—our drugs.

In June 2012 came the announcement that scientists had finished genetically sequencing the roughly eight million genes in ten thousand species of microbes that were found dwelling inside 242 healthy human bodies—the first fairly comprehensive map of the human "microbiome." A significant number of diseases had previously been linked to combinations of genes found in humans and genes found in our microbial guests. The systematic analysis of the combined human-microbe genome will now lead to new strategies for maintaining health and preventing infections, and to better treatments of metabolic disorders and various diseases of the skin and digestive tract.

Other microbiologists, according to one recent count, have already sequenced 1,554 bacteria, 2,675 viral species, some 40,000 strains of flu viruses alone, and more than 300,000 strains of HIV. Each genome, in the words of Dr. David A. Relman, a professor of medicine, microbiology, and immunology at Stanford, provides a "master blueprint of a microbe"—it is "like being given the operating manual for your car after you have been trying to trouble-shoot a problem with it for some time."

The manuals will require constant updates because germs are quick-change artists. When stressed by drugs or other environmental pressures, bacteria shut down their own repair machinery to mutate faster. They exchange genes in ways never imagined until the recent arrival of gene-tracking tools. The definitive diagnosis of the disease killing the individual patient in *this* hospital bed is the germ's genome—which gets sequenced when the "tuberculosis" or "syphilis" (two conventional, one-size-fits-all names given to two familiar clusters of clinical symptoms) unexpectedly fails to respond to the standard antibiotics.

By analyzing large numbers of drug-resistant germs, researchers can then identify the genes that confer the immunity. The NDM-1 gene protects *E. coli* against almost all current antibiotics. Four genes are known to provide resistance to tetracycline or sulfonamide. Friendly bacteria commonly found in the human gut carry hundreds of genes that confer varying degrees of resistance to thirteen antibiotics. Some genes go a step further, allowing microbes to feed on the debris of the drug they have learned to dismantle, and even become what have been called "bacterial drug addicts, unable to survive without the poison given to exterminate them." Understanding the chemistry that confers resistance will help guide the development of new antibiotics that can dodge it.

Today's biochemists can also get out ahead of Darwin and design antidotes to germs that don't yet exist. By analyzing a protein that made the 1914 flu virus particularly lethal, for example, designers of flu vaccines have significantly improved their ability to stay ahead of one mutable enzyme that plays a key role in the viral chemistry. They are working on a long-term flu vaccine, assembled from an array of different proteins that have popped up in the past.

More commonly, doctors assemble complex drug regimens by hand. Some microbes mutate so fast that the antidote cocktails have to be tailored to individual patients and adjusted on the fly. That is why medicine has sequenced some three hundred thousand strains of HIV. By combining genomic and other data with clinical records of how tens of thousands of patients fared on different drug cocktails, a powerful digital engine (discussed further in Chapter 7) predicts future patient outcomes better than do most doctors who specialize in the field.

———————

THE TREATMENT OF rogue human chemistry is on the same trajectory. Here, too, biochemists are fast acquiring catalogues of the human genes, proteins, and other molecules that—when corrupted, remixed, or switched on or off—play a role in propelling a biochemically distinct disease. But with most complex diseases, we often don't yet know how the molecular building blocks combine to evade the drugs we already have. Nor do we know which molecules we should target to beat each combination when they do.

Much of the time, the only practical way to find out is to develop a wide array of drugs that target the molecular building blocks, and let doctors who specialize in particular diseases work out how to combine them, one patient at a time, to provide effective treatment. Guided by molecular analyses of the molecular structure of the disease, they assemble drug cocktails on the strength of biochemical logic and clinical experience. The accumulation of this experience leads medicine to underlying biochemical patterns that some patients have in common. This is the process that culminates in what oncologists now call treatment algorithms—sets of rules for assembling available drugs into cocktails to fit the disease in the individual patient.

Drug designers are also working on a new generation of vaccines that persuade immune systems to launch multipronged attacks on patients' own rogue chemistry. "Cancer immunotherapy comes of age," *Nature* magazine recently announced. As of late 2011 vaccines aimed directly at fourteen different types of cancer were undergoing clinical trials. One company is developing a drug that consists of a cocktail of common markers found only on cancer cells, together with drugs that boost immune system activity. That company was focusing its research on melanoma and lung cancer, but the same markers show up on liver, bladder, and many other types of cancer cells, and early trials suggest that genetic screening of a tumor can determine whether the vaccine is likely to help beat it.

In early 2012 Mayo Clinic researchers reported promising results in mouse tests of a skin cancer vaccine that consists of a cousin of the rabies virus, genetically modified to include a broad range of genes from cancerous tissue. The team also has liver cancer vaccine trials under way and is exploring vaccines for more aggressive lung, brain, and pancreatic cancers. "I do believe we can create vaccines that will knock them off one by one," declared one of the researchers. "By vaccinating against multiple proteins at once, we hope that we will be able to treat both the primary tumor and also protect against recurrence."

Other researchers are working on "negative vaccination" for autoimmune diseases. The trick here is to teach the patient's immune system to tolerate what it will otherwise end up attacking. Type 1 diabetes, for example, is caused by an autoimmune attack on various proteins found on the surface of insulin-producing cells in the pancreas. Researchers are now pursuing vaccines that expose a patient to one or more of the proteins, or to insulin itself, because it may be involved in triggering the attack. A similar approach, using three protein segments thought to trigger celiac disease, is also being explored.

As experience accumulates, the analysis of databases that combine clinical experience with cocktail therapies and patient profiles will reveal any simpler molecular patterns that may be lurking under the complexity. Medicine now has the tools to analyze the biochemical structure of a disease at almost every stage of its development, and thus can quickly develop reliable algorithms for using simpler cocktails or just a single drug when appropriate. The vaccines designed to treat late-stage tumors may also be usable much earlier. Women who carry especially hostile combinations of breast cancer genes, for instance, now start taking a cocktail of estrogen-blocking drugs prophylactically. A combination of two HIV drugs taken prophylactically—lightweight medicine compared to treatments required at later stages of the disease—cuts the risk of infection in high-risk groups roughly in half.

But given the endless variety of genetic SNPs and copy number variations scattered across every human body, tracking diseases back to their precise genetic origins may reveal instead the need for an endless variety of biochemically distinct drugs to treat what has traditionally been viewed as a single disorder. Prostate cancer is a good example. In 2010, the FDA approved the first true cancer vaccine, Provenge, to treat it—or more precisely, approved a process for preparing a unique vaccine for each patient. White blood cells extracted from the patient are fused with a protein that combines an antigen found on prostate cancer cells and a signaling factor that helps the cells mature. Returned to the patient, the cells trigger an immune response against the cancer. In August 2011 a team at the University of Pennsylvania that had used a similar process of using the patient's own genes to reengineer white cells culled from his own body reported "sensational" results in several patients suffering from the most common form of leukemia.

IT IS BY ignoring clinical symptoms and human crowds and descending into the molecular realm that today's biochemists, doctors, and the free market now rediscover and once again capture the full power and economies of socialized medicine. Nucleic acids and thus the proteins that they define are contagious (in humans, sexually transmitted) progenitors of health and disease—they flow through the river of life as surely as cholera once flowed through London's water supply. Mapping human chemistry exposes the differences among many forms of breast cancer—but it also exposes the shared biochemistry of leukemia and gastrointestinal cancer. There are differences everywhere, but there are also matches, overlaps, and widely shared forms of biochemical strength and weakness.

Going forward, molecular science will link more drugs to more genes, proteins, and other molecular constituents of the sick and healthy parts of our bodies. The more we learn, the easier it will get to learn still more, and the cheaper it will get to translate what we know into powerful medicine. The progressive accumulation of increasingly detailed descriptions of how biochemical networks function will progressively lower the cost of designing new drugs and determining quickly and cheaply how they can be prescribed safely and effectively.

Databases will expand to include the results of laboratory experiments with microbes and test animals, and thus expose biochemical webs and processes that operate in the same way in different species. No bacterium, rat, monkey, or other animal is a good model for an entire person, but some are enough like us in ways that are relevant to beating a particular disease, and they are now bioengineered to incorporate human—and sometimes the individual patient's—immune system or cancerous tumor genes. The discovery of a good animal model for a human disease is often what launches the ultimately successful search for a drug to beat the disease in people. Extended across species, molecular cartography can thus steadily improve our ability to develop solid human-drug science outside human bodies.

Medicine will know that it has reached the pinnacle of drug science when it is able to design a new drug and predict how it will perform well in which patients without first testing the drug in even a single patient. Two decades ago, the suggestion that this might someday be possible would have been dismissed out of hand. With the right regulatory and economic policies in place, we might well get there within a few decades. We know how to design precisely targeted drugs. We have the tools to create vast databases of molecular and clinical data that span human bodies, the microbes they host, and a limitless number of drugs. And we have the ever expanding power of our digital technology to work out how the pieces interact.

Part Two

SCIENTIFIC MEDICINE

6

THE SOCIAL CONTRACT

WE ARE NOW over a century into a legal struggle that has yet to end. It concerns the patient's right to control his or her own body—all of it, not just the procreative parts—and the drug company's right to supply the drug that perhaps will provide the control. The "perhaps" is at the center of the struggle: who decides whether it will or won't?

In 1902, the Constitution affirmed the patient's right to make all the calls. Today a blend of constitutional and statutory law affirms Washington's authority to decide whether a medicinal drug will provide the control that the patient seeks, while also affirming the patient's right to use any drug, once licensed, pretty much as the patient and doctor see fit. In 1902, private contract law controlled what, if anything, the vendor of a drug promised, and thus sharply limited how often tort law and juries became involved in passing judgment on the drug's performance. Today, tort law rules, and juries are free to second-guess Washington's views on what drugs do or don't cause.

Here, then, is the schismatic legal framework within which biochemists and doctors now struggle to develop modern drug science. The drug company owns the drug and is responsible for all the explicit and implicit legal promises that come with it. The patient and doctor own and control the biochemical environment that spawns and incubates the disease that the drug is supposed to cure, along with all the bystander chemistry that might be involved in side effects. FDA regulations and, indirectly, the open-ended threat of lawsuits strictly limit interaction between the designers and users of drugs.

So the law often finds itself at war with itself—it demands good drug science while severely obstructing the collaborative process that is now an integral part of developing that science. For the last twenty years, as discussed in the next two chapters, Washington has nudged the law this way and that, responding to medical crises and shifts in the political winds. Washington has already tried major parts of the reforms that are now needed, and many of those parts remain on the books. But to understand their importance, and why they should be reinvigorated and applied more systematically, we must begin with the tangled and often self-contradictory status quo, and how deeply embedded it is in the federal drug law, constitutional jurisprudence, and the tort system.

———————

ASSISTED BY MASS advertising made possible by the rise of newspapers, unscrupulous peddlers of quack nostrums proliferated in late nineteenth-century America. By 1902, the rest of Washington was itching to intervene, but the Supreme Court wouldn't let it. The Constitution secured every citizen's right "to enter into all contracts which may be proper, necessary, and essential" to getting on with private life, and that certainly included private health. Washington could go after fraud—deliberate attempts to mislead—but honest ignorance was a complete defense. To be doubly safe, the peddler of a quack cure could protect himself with a formulaic disclaimer regarding all the things that he hadn't investigated, didn't know, and wasn't promising.

J. H. Kelly was the proud inventor of a "practical scientific system" of magnets to transmit cures directly from his Missouri offices into the bodies of patients across the country. He was raking in more than $1,000 a day when his local postmaster stopped delivering his mail and cashing his money orders. A federal law empowered the postmaster general, "upon evidence satisfactory to him," to cut the mail out of mail-order fraud. The postmaster general was satisfied that Kelly's magnets weren't curing anyone.

Supreme Court justice Rufus Wheeler Peckham wasn't so sure. "Ability to cure" was a matter of opinion, not fact, he reasoned. Kelly had never claimed his cure would always succeed. In medical matters, there was "no exact standard of absolute truth by which to prove the assertion false and a fraud." Under the Constitution as then interpreted, that took Washington out of the picture.

The Federal Food and Drugs Act of 1906, signed by Teddy Roosevelt, suffered the same fate in the same courtroom, allowing one Dr. Johnson to continue peddling his "Mild Combination Treatment for Cancer." The law prohibited all "misleading" claims about a drug's properties. Making clear that any broader reading of the law would render it unconstitutional, Justice Oliver Wendell Holmes read it to apply only to "plain matter[s] of fact"—claims about the contents of the bottle, for example—not any "estimate or prophecy concerning their effect." A chastened Congress replaced "misleading" with "fraudulent." The Dr. Johnsons couldn't be touched so long as they kept any promises about the drug's effects too vague to mean anything much at all.

There are no fraudulent opinions, but there are limits to how much shelter hucksters can find in cleverly asserting their belief in falsehoods. So, working with Washington's Bureau of Chemistry, the postmaster general hired a full-time scientist to analyze quack medicines, and as germ-killing science improved, judges became increasingly willing to call quackery fraud. In 1916, one judge concluded that expert opinion about tuberculosis and its antidotes was too divided to prove that the claims made about Tuberclecide were knowingly false—but by 1928, another court was ready to side with the postmaster.

Good-faith ignorance could still shelter a lot. In 1937, Harold Watkins, chief chemist for the Massengill drug company, set out to improve the taste of one of the new sulfa drugs. He dissolved the drug, along with raspberry flavor, caramel, and amaranth to color it reddish purple, in diethylene glycol, otherwise known as antifreeze. "Elixir Sulfanilimide" promptly killed 105 people, including thirty-four young children, spread across the country from Virginia to California. Watkins shot himself in the heart with a .38-caliber pistol. Massengill paid a nominal $26,100 fine for "misbranding," because the word *elixir* falsely implied that alcohol had been used as the solvent. So far as Washington was concerned, Massengill's only mistake was its failure to blend some honest liquor in with the antifreeze.

———

WHILE MASSENGILL WAS bottling antifreeze, the New Deal Supreme Court, in 1937 cases upholding the constitutionality of labor union and minimum-wage legislation, had concluded that four decades of libertarian constitutional jurisprudence had all been a big mistake. In 1938, Congress enacted a new drug law: thereafter, no drug could be sold until the manufacturer had established, to Washington's satisfaction, that the drug was safe. The FDA then announced that the label on the bottle had to describe what the drug would supposedly cure. Factual claims about "ability to cure" were now mandatory, and Washington had broad discretion not to license a drug until it was persuaded that they were true.

Federal drug law thus crossed its Rubicon in 1938. As so often happens in Washington, however, nobody knew it at the time. The changes seemed small and sensible, and they were, given the science of the time. The front-end licensing system established in 1938 reflected the simple, elegant vision of molecular medicine that had won Paul Ehrlich a Nobel Prize in 1908. The drug's story is the disease's, told in reverse. A single, straight line links the drug to the root cause of the disease and the patient's return to health. A good drug hits its intended target and nothing else.

Between 1938 and the early 1970s, the period during which Washington transformed that idea into a large compendium of rules and protocols, deciding which bullets really were magical was quite easy, or so it appeared. Diseases were diagnosed late, and so far as medicine knew, one person's chemistry was much like another's, so a new drug didn't need to be tested for very long, nor did trials have to involve many patients. Clinical trials would track clinical symptoms because they were what mattered to patients.

In this state of ignorance, it was all too easy for well-meaning people in Washington to conclude that they could manage drug science from the top down, seeing to it that all the important questions were resolved under Washington's supervision before a drug was unleashed in the marketplace. Every drug would be sold with a compact, Washington-approved label that would tell doctors and patients under what conditions it would perform safely and effectively.

The main worry was that wishful thinking by the clinicians and patients involved in the trials would bias the results, allowing bad medicine to continue slipping through the cracks. The scientific community, however, had already worked out how to prevent that kind of error. In 1938, the U.S. Public Health Service tested a pertussis vaccine in Norfolk, Virginia, in what is thought to have been the first randomized, double-blind trial of any pharmaceutical product. Eight years later, British researchers conducted what may have been the first double-blind placebo-controlled trial of a curative drug, streptomycin. The FDA followed their lead, and these were the protocols that would soon be adopted as the "gold standard" for modern drug testing. But the protocols lead to what can, at best, be called crowd-science medicine—though, anchored as they are in blind statistical correlations, they are almost all crowd and very little science.

The statutory framework of our modern drug law was put in place after the world learned a particularly brutal lesson about the perils of drugs that tinker directly with human chemistry. Synthesized in the early 1950s, thalidomide was a gentle, effective sleeping pill that was specifically recommended for use by pregnant women suffering from morning sickness. The drug also had a diabolical power to survive the acid in a mother's stomach, slip through the lining of her gut, glide past the protective biochemical sentinels in her liver and through the protective barrier of a placenta, and instruct the baby in her womb to stop growing ears, arms, or legs. Thousands of thalidomide babies were born in western Europe and elsewhere.

Luckily for America, Dr. Frances Kelsey, a newly hired FDA medical officer, had kept thalidomide out of the U.S. market by stretching out her review of the license application. Thalidomide was the first application to reach her desk, and she wasn't satisfied that the drug had been adequately tested—some users had experienced painful tingling in their arms and legs. "We were dickering with them—do more studies, label it differently," she recalled in a 1994 interview. The horror of the thalidomide disaster in Europe did, however, help propel major amendments to the 1938 drug law through Congress, and President Kennedy signed them into law in late 1962.

The amendments expressly codified, for the first time, a requirement of up-front proof that a new drug really worked, but in fact the FDA already had the authority to require that by mandating and regulating efficacy language on the label. Their main practical effect was to launch the drafting of rules that would spell out how Washington would oversee the development of molecular medicine and decide whether it was good enough for public consumption.

Discussions of the 1962 law invariably mention its "safe" and "effective" language but often omit the all-important qualifier: that Washington certifies the drug's future safety and efficacy only "under the conditions of use prescribed, recommended, or suggested in the . . . labeling thereof." The label, in other words, is where the license addresses the patient-side factors that must be present or absent for the drug to perform as promised—the disorder to be treated, along with other drugs that the patient may be using, the patient's age, alcohol consumption, pregnancy, and other factors that shape the biochemical environment in which the drug will operate. The 1962 law demanded "substantial" evidence, derived from "adequate and well-controlled" clinical trials. Thus no drug is safe or effective until the FDA concludes that a sufficient body of scientific evidence has been amassed and collapsed onto a label that will allow doctors to prescribe the drug to the right patients. The 1962 law also gave drug companies two years to come up with acceptable proof that some four thousand drugs licensed before 1962 met the new requirements.

The FDA took eight years to finalize the rules that defined what it would view as substantial, adequate, and well-controlled—and made clear that it would summarily reject any application based on studies that hadn't precisely tracked the approved protocols. Washington spent the next forty years enforcing, expanding, contracting, and tinkering with those rules, in a febrile, politically fractious process that continues to this day.

While the rules were being written, biochemical science was developing the tools that would expose our complex biochemical diversity and revolutionize drug design. But the benefits of those drugs came later. For safety regulators, the new focus on molecular-scale factors made it easier to imagine how biochemical tinkering might spawn cancer or birth defects. In the early 1970s, with memories of thalidomide still fresh, statistical studies linked strokes and several forms of cancer to the different forms of estrogen in birth control pills and DES, a widely prescribed synthetic estrogen (incorrectly) believed to reduce the risk of miscarriage. In 1970, Congress held inflammatory hearings, blaming inadequate FDA oversight of estrogen-based birth control pills for a litany of nasty side effects, some real, many not. The FDA grew increasingly worried that small, quick trials would miss rare side effects that developed slowly. In dealing with vaccines, contraceptives, and other drugs that would be widely used by healthy people, the emerging safety-first mind-set meant conducting huge, lengthy trials.

Then concerns about diversity and fairness entered the picture. Though the motives for including them were more political than medical, they made rough scientific sense, too. If patients are visibly different, they must also be somewhat biochemically different. Thus six times between 1988 and 2002 Washington demanded more diverse representation of racial, ethnic, gender, age-related, and other "population subgroups" in clinical trials. As recently as 2005 the FDA licensed BiDil, a mixture of two older drugs for treating heart disease in "self-identified black patients"; a few months later, the agency directed drug companies in general to analyze clinical data using the six sex/race/ethnicity categories defined by the Office of Management and Budget to enforce civil rights laws in education. As the FDA sheepishly acknowledged at the time, those categories are "sociocultural construct[s]," not science. The patient, not the doctor, decides which box to check, and he may check more than one.

In a 1994 directive explaining how much diversity must be tested in trials that they fund, the National Institutes of Health (NIH) added that "cost is not an acceptable reason for exclusion." For drug companies, however, the cost of inclusion became a compelling reason to pursue only those drugs that could be sold to lots of patients once licensed—which gave the FDA still more reason to require even bigger, longer trials. Arm in arm, the FDA and its wards have tracked diversity into a quagmire of human trials that never stop growing, take forever, and cost the earth.

WHEN THE SUPREME Court upheld Kelly's right to sell magnetic healing, it was reaffirming the general "liberty of contract" it had discovered in the Constitution five years earlier. The word *liberty*, Justice Peckham had explained for the unanimous Court, secures every citizen's right not only to "be free from the mere physical restraint of his person, as by incarceration," but also to contract with others to buy or sell products and services pertaining to the "enjoyment of all his faculties." Contracts involve two parties, and in affirming Kelly's right to sell a cure, the Court affirmed his customers' right to buy it. By grounding the right in private contracts, the Court upheld the patient's private right not only to control his own body but also to choose how best to control it—so long as the consequences of the choice were altogether private. The Court soon found itself deciding a case in which they weren't.

In October 1901 diphtheria, at that time the most common killer of American teenagers, swept through St. Louis. Not long after, Emil von Behring collected the first Nobel Prize in Medicine for his development of the first diphtheria serum—antibodies collected from the blood of vaccinated horses. The timing was unfortunate. A horse named Jim, previously the hauler of a milk wagon, had been enlisted to manufacture serum destined for St. Louis. Contaminated with tetanus, the Jim-Behring serum killed five-year-old Veronica Neill, along with a dozen other children. At about the same time, nine other children in Camden, New Jersey, were killed by tainted smallpox vaccine.

In response, Congress passed the Biologics Control Act, which Teddy Roosevelt signed in July 1902. Today that act, which covers medicines derived from live sources, is often described as America's (and possibly the world's) first drug licensing law. But it wasn't, and probably wouldn't have survived a constitutional challenge if it had been. A few months after its enactment, Justice Peckham, in his opinion addressing Kelly's magnets, made a point of mentioning that the smallpox vaccine's "ability to cure" was likewise a matter of opinion, not fact.

As originally enacted, the Biologics Control Act licensed manufacturers, not their products. The standards promulgated by the government's Hygienic Laboratory, which administered the new law, focused on how germs were cultivated and then transformed into vaccines, and how facilities should operate so as to avoid risks of contamination. (The lab would later be renamed the Laboratory of Biologics Control and is now the Office of Biologics Research and Review.)

In enforcing those standards, vaccines were examined for "purity and potency," but manufacturers remained free to promise nothing at all about how the vaccine might affect the patient. The law required an accurate label on every biologic package or container, but only to identify the product and its manufacturer. Many labs were being managed horribly, and the new standards quickly drove them out of the market. When the dust settled, only about ten remained.

Two weeks after President Roosevelt signed the Biologics Control Act, there was an outbreak of smallpox in Cambridge, Massachusetts. The city's Board of Health ordered everyone to get vaccinated. Henning Jacobson refused, and presented the Peckham defense. As summarized by the U.S. Supreme Court, Jacobson's argument was that vaccines were often "impure and dangerous," and it was "'impossible' to tell 'in any particular case' . . . whether [a vaccine] would injure the health or result in death."

Cambridge had no right to decide what kind of medicine was good or bad for his personal health. Mandatory vaccination, Jacobson argued, was unconstitutional.

Over Justice Peckham's dissent, the Supreme Court disagreed. Cambridge wasn't trying to protect Jacobson; it was trying to protect other residents from being infected by Jacobson. The Board of Health had left open the possibility of allowing residents to demonstrate why they couldn't safely be vaccinated, and doctors did have views on who those people might be. But it was up to Jacobson to show why his body couldn't tolerate the vaccine, and he hadn't even tried.

However misguided it may seem to those of us today who gladly trust Washington to protect us from the junk cures and see to it that we can, if we wish, get the good ones, a constitutional right to pay good money for magnetic healing that isn't also a right to reject a free jab of *vaccinia* reflects a coherent view of where the right begins and ends. So does a Constitution that accepts Washington's power to regulate the manufacture of Jenner's vaccine but not the sale of magnetic cures. The Supreme Court sided with the government when the solid science of that era pointed to an indubitably public problem—the risk that contaminated vaccines or human bodies would spread infectious disease. But until the medical science was able to establish how medicine was going to affect any individual patient, the authorities had no constitutional right to substitute their own uncertain medical judgments for the patient's in matters that affected his body alone.

———

JACOBSON WOULD HAVE no constitutional right to buy magnetic cures from J. H. Kelly today, even though his private health alone was at stake. Jacobson's right to remain unvaccinated, on the other hand, would be upheld in much of today's America. Many states have laws that expressly protect it, and today's Supreme Court might well conclude that the Constitution does, too. Today's Constitution affirms the individual's right to choose even when his choices help spread infectious diseases and breed drug-resistant germs. At the same time, however, our Constitution now accepts a licensing regime under which every new drug, however private its effects, is presumed to threaten public health. When drugs are involved, private choice itself is now viewed as perilous; no patient may use a drug until the FDA concludes that it will be both safe and effective in a very large majority of future users. A drug's "ability to cure" any individual patient is of no interest other than as a data point to use in calculating the drug's impact on the statistical health of the crowd.

The Supreme Court opinion upholding Jacobson's right to remain unvaccinated would cite a 1965 decision in which the Court struck down a law that could have been challenged and might well have been struck down in Justice Peckham's courtroom, too—an 1879 Connecticut law, written by P. T. Barnum before he got into circuses, that made it a crime to prescribe any drug to prevent pregnancy.

In late 1961, Estelle Griswold, a Yale Medical School professor and a top official of Planned Parenthood, had decided to challenge the law, and started prescribing Enovid, the first mass-market birth control pill, at a clinic she set up in New Haven, Connecticut. She was arrested, convicted, and fined $100.

In four separate opinions, seven Supreme Court justices agreed that the Connecticut law was unconstitutional. Justice William O. Douglas wrote the lead opinion for a four-justice plurality. Medical aspects of procreation, he declared, are sheltered from government meddling by a nebulous "zone of privacy" in the Bill of Rights, a zone he discerned in the "penumbras" of the First, Third, Fourth, Fifth, and Ninth Amendments.* Their shadows have been spreading across the legal and cultural landscape ever since. Today, *Griswold*'s legal progeny constitute a broad body of what legal scholars often call the law of "privacy and personhood," or, more colloquially, the "right to control your own body."

Conspicuous by its absence in *Griswold* itself was any mention of the old "liberty of contract." Indeed, Justice Douglas made clear that the *Griswold* right of 1965, unlike the *Kelly* right of 1902, covered only the right to prescribe and use the pill, not the right to design and sell it. Washington remained free to regulate "economic problems, business affairs, or social conditions." The pachyderm lurking in the corner of the courtroom was, of course, thalidomide and the new drug law that President Kennedy had signed three years earlier. Justice Douglas didn't want anyone to suppose that Dr. Griswold's right to prescribe Enovid, and her patient's right to buy it, placed any serious limits on the FDA's power to decide whether the pill belonged on the market at all. Eight years later Justice Douglas made that doubly clear when the Court, ruling on a challenge brought by the manufacturer of a drug that was supposed to prevent miscarriages rather than pregnancy, upheld rules the FDA had written to implement the 1962 drug law. Washington "surely has great leeway in setting standards for releasing on the public, drugs which may well be miracles or, on the other hand, merely easy money-making schemes through use of fraudulent articles labeled in mysterious scientific dress," he declared.

---

*Those clauses protect freedom of speech and association, and they affirm your rights not to have soldiers quartered in your home in peacetime, not to be subjected to any unreasonable "search" or "seizure" of your "houses, papers, and effects," and your right to remain silent, along with other unspecified rights "retained by the people."

Clear but not coherent. A private right to control your own body can't be squared with a great-leeway government power to decide what's good for it. This much, at least, Justice Peckham got right—a right to control your own body implies a right to buy what you need to control it. Justice Douglas just tried to bluff his way through.

Justice Peckham also addressed the issue head-on and drew a line that could move as the advance of drug science transformed the peddling of preposterous opinions into fraud. By the time Justice Douglas got involved, however, it was settled that until Washington certified it, there was no such thing as drug science, or at least none good enough to be bottled and sold. The Biologics Control Act had been rewritten in 1944 to include the licensing of biological drugs—thereafter, manufacturers had to establish that their products were "pure" and "potent" but also "safe." Manufacturers of all other drugs had to establish that they were "safe" and "effective." At the threshold, a drug's "ability to cure" was *never* something that the patient or doctor might be able to judge better than Washington could.

And yet that was obviously false, even in 1965, and it certainly isn't true today. The notion that a political act is what makes science reliable is one worthy of Lysenko and the master he served. Real scientists stopped saying such things around the time of Galileo. The thalidomide drug law merely placed old and familiar testing protocols under Washington's tight control.

Those protocols, moreover, were all anchored in the study of crowds, not individuals. In 1969, the FDA acknowledged a link between birth control pills and fatal blood-clotting disorders but decided that the pill was nonetheless still safe enough. "Safe" isn't a scientific term, and the process of balancing one woman's stroke against the sexual autonomy of a thousand others is no more scientific than Kelly's magnets. The safety screw, as we shall see, has been officially loosened to accelerate the approval of some drugs, and it can easily be tightened to keep other drugs off the market. The FDA has recently found itself flipping as fast as any politician can flop over the possibility that giving teenage girls over-the-counter access to a week-after pill might promote risky promiscuity or be hazardous in other ways. (For the details, see Chapter 13.)

A constitutional right to control a quintessentially private aspect of personal health is meaningless without a right to make a private call about how to balance the risks and benefits of using the product that provides the control. Constitutional rights belong to individuals, not crowds. And since 1965, medical science has moved decisively away from the crowd and into the molecular depths of the individual patient.

As that has become increasingly clear, patients have repeatedly challenged the constitutionality of the FDA's licensing power on *Griswold* grounds—and have lost every time. That power is constitutional, courts have concluded, even when it stands between a terminally ill young cancer patient and the drug she desperately seeks—a drug, one might add, that was licensed three years after she died. At the same time, as we shall see, *Griswold*'s sketchy recognition of a private right to control one's own body has been reaffirmed and expanded in subsequent Supreme Court rulings and codified in a wide range of federal and state laws that address privacy and discrimination. We have thus arrived at a world in which the individual has a broad private right to control his or her own body, except when the control might be supplied by a drug. You personally have no right to use a drug until the FDA has determined that it is safe and effective for the statistical crowd.

That gets us to the glaring self-contradiction in the 1962 drug law. Doctors remain free to prescribe licensed drugs "off label." Patients remain free to trust their doctor's prescription and ignore Washington's. The patient may also live an off-label life, smoking while on oral contraceptives, washing down sedatives with vodka, and mixing uppers and downers. The license's implied promise of efficacy is, of course, meaningless without a specific effect in mind. And many potent drugs are in fact toxic, too; they are called "safe" only when they are controlling an even more dangerous disease. The label is the Washington-approved attempt to license patients, but feel free to ignore it.

———————

WHETHER THE LABEL is ignored or followed to the letter, the patient may sue the drug company if things don't work out as hoped. In Justice Peckham's world, the law of private contract decided who won. The vendor rarely promised anything all that specific, so juries rarely had a chance to conclude that magnets or mild cancer cures were worthless or worse. Half a century later, however, the federal government's own Biologics Lab played a central role in choreographing a disaster that launched a second, sweeping repudiation of contract law—this time in favor of tort law.

In April 1955 five-year-old Anne Gottsdanker's legs were left paralyzed by a shot of Salk's new polio vaccine, supplied by the Cutter Company, that contained traces of the live virus. It caused mild forms of polio in some forty thousand children. About two hundred were left paralyzed. Ten died.

The March of Dimes had funded the development of Salk's vaccine and overseen the first large field trials. The Biologics Lab had quickly licensed five private companies to produce it for commercial sale, specifying how it was to be manufactured. Its specifications were based on those that had been developed by Salk and tested by two large, experienced manufacturers that supplied the vaccine for use in the trials that got the vaccine licensed.

Cutter, the smallest of the five licensees, had followed Washington's instructions to the letter. But the lab, Washington's anointed guardian of vaccine science, had botched its rewrite of Salk's protocols. As a former employee would recall, the lab just "didn't have the expertise to handle it."

Cutter had not been negligent, the jury in Anne's case concluded. But the judge had instructed the jury that Cutter could, alternatively, be held liable on the ground that the vaccine came with an "implied warranty" that it contained no live virus. Having breached a promise written in print too fine for anyone but a judge to read, Cutter was responsible for Anne's injury.

This was old contract law dressed up in new legal jargon—and judges soon discovered that it wasn't quite what was needed. Implied promises can be expressly disclaimed, which is just what drug companies started to do. When vaccine companies were accused of harming patients in subsequent cases, the manufacturers had invariably conveyed exactly what Washington had told them to convey, to the doctors and medical societies whose responsibility it was to convey that information to patients in terms they could understand. Judges then announced that in mass vaccination programs where nurses handled most of the needles, the drug company had to take steps to inform patients directly about all the reasons they might prefer not to be vaccinated. The opinions in these cases affirm the patient's right to an "individualized medical judgment" so that he may exercise his private right not to do what the Board of Health is imploring him to do.

Meanwhile, other judges had decided that no amount of paper, warning, or doctor's advice would ever be dispositive—implied promises had given way to "strict liability." Judges and juries would be free to conclude that a vaccine was "defective," "unfit" for its intended use, or "unreasonably dangerous," either "as designed" or "as marketed." In applying these standards, jurors are instructed to focus on what's best for the crowd, not the individual patient.

Not long after the Cutter tragedy, the Biologics Lab licensed Albert Sabin's polio vaccine, which used live but weakened strains of the polio virus delivered on a sugar cube. Sabin's vaccine was slightly riskier than Salk's but more broadly effective. Patients preferred sugar to a needle, and because the live vaccine was contagious, they went home to vaccinate about one-quarter of the people with whom they came in close contact. Washington understood the trade-offs and promoted Sabin's vaccine in preference to Salk's.

The ensuing Sabin vaccine lawsuits pointed to a far graver problem that would, in the next two decades, come perilously close to shutting down the vaccine industry completely. In March 1963 Glynn Davis, thirty-nine years old, swallowed his cube and thirty days later found himself paralyzed from the waist down. Wyeth, the company that had manufactured the vaccine, had been scrupulously careful in cultivating the seed virus, testing the vaccine, clearing it with Washington's Biologics Lab, and shipping it with all the required warnings. A jury found no breach of any implied promise, but a federal appellate court ordered a retrial because Wyeth hadn't directly warned Davis himself.

In a footnote, the court mentioned that the residents of West Yellowstone, where Davis lived, had been vaccinated because a wild polio virus was on the loose in the community, and that there was, at the time, "no method of knowing with medical certainty" whether Davis had been crippled by nature's virus or Wyeth's. That, of course, created a risk that the vaccine would be blamed for causing a disease that it was in fact preventing. Or, in some future case, incorrectly blamed for having caused some completely unrelated harm. In another polio vaccine case decided six years later, Wyeth wasn't even allowed to present the testimony of a virologist who had concluded that a wild strain of the virus had caused the patient's polio.

Such rulings did often transfer money from rich drug companies to poor and frequently very sympathetic young plaintiffs, and were rationalized by some legal scholars as providing an economically efficient form of insurance to all. But the high and unpredictable stakes eventually helped drive most companies out of the vaccine market.

––––––––

THUS WE ARRIVED at a market for drugs that operates under a social contract, written by agencies, expert committees, judges, and juries. This muddled and often self-contradictory legal regime makes the development of drug science far slower, costlier, and more unpredictable than it could be, quite often to the point where the private sector won't develop it at all. It preempts the only legal instrument—the private contract—that allows private parties to assign risks and responsibilities in clear, definite ways that reflect who knows, or has access to, the information needed to develop the science at the heart of today's molecular medicine, transform it into drugs, and prescribe them well.

The social contract assumes that drug companies are close to omniscient, or could be if only they took their responsibilities seriously. The drug company has to work out the drug science up front, at its own cost, and keep working until Washington is satisfied that enough has been learned to compose a label that will allow the drug to be prescribed safely and effectively—fuzzy terms that imply much but have no firm anchor in hard science. Thereafter, drug companies may communicate and collaborate with doctors and patients only under rigid rules, protocols, and scripts that Washington has approved.

The front-end trials are conducted by doctors working under Washington-approved protocols, with blindfolds on. Everything that they don't yet know about the molecular details that affect the drug-patient interactions—that estrogen, for example, accelerates some breast cancers and slows down others, depending on the presence or absence of estrogen receptors on the cancer cells—is supposedly covered by the front-end process of selecting patients to include in the trial and the statistics used to evaluate the drug when the trial ends. Learning and adapting as you go—an integral part of what good doctors now do when treating complex disorders out in the real world—isn't part of the process.

The responsibilities of Washington's science police expand as fast as biochemists uncover new details about our biochemical diversity and the complexity of the interactions between drugs and people, and they expand further every time a drug company develops a new drug to add to the stew. For trial lawyers, the new science offers endless, lucrative opportunities to sue. Claims can culminate in jackpot verdicts and runaway class actions that make J. H. Kelly look like a small-time piker.

The overarching message from the government to the patient is this: *Trust us. Washington has the drug science covered, so you don't have to worry about it.* But as with many messages from Washington, there's much dodgy detail under the arch. *Consult your doctor* is one of the details; *he knows best. And if it turns out that he didn't, consult your lawyer, who will consult a jury—the jury has the last word.* Washington's social contract makes vague, sweeping promises, assigns no final responsibility, and assumes none itself.

Washington can declare the drug safe and effective; a jury can declare it so dangerous and defective it doesn't belong on the market at all. The FDA minutely controls the warnings supplied with every drug and, to avoid scaring patients off their medicines, forbids too much warning as strictly as too little. Too little may still be the plaintiff's central claim in a subsequent lawsuit. So one arm of government limits what the drug company may say, while another invites juries to read lies or broken promises into its silence. Further, California's courts aren't bound by New York's, and tomorrow's jury isn't bound by yesterday's.

Finally, a drug can be judged unfit for the market because it harms some patients in ways that only juries can discern. Washington strains to work out all the science up front; juries work it out years, decades, or sometimes even a generation or two later—the liability tail can stretch out almost indefinitely. Expert witnesses routinely peddle to juries data and conclusions that would be instantly rejected if Merck presented them to the FDA.

This is not an environment conducive to the orderly development of the complex disease-science and drug prescription algorithms that now guide the treatment of, say, breast cancer. But then again, much of the breast cancer science was developed well outside the old legal framework, after Washington was suddenly smacked in the face by biochemical reality.

The expansion of the private right first articulated in *Griswold* played a role in delivering the smack. And while the suggestion will too easily be branded as a reactionary proposal to return to the days when drug companies killed children with impunity, here's where this part of the story will end several chapters hence. The advance of molecular medicine will soon grind to a halt if we don't return to a modern-day version of a legal regime much like the one that not only prevailed but was a matter of constitutional right in 1902. "Ability to cure" is a matter of fact, not opinion. But most of the relevant facts are buried deep inside patients' bodies and mirror the complexity and diversity of human chemistry. As currently interpreted and enforced, the thalidomide drug law and the *Griswold* Constitution impede rather than promote the development of the complex science that leads to the cures that work.

7

A VIRUS LIKE US

So WHAT WOULD a watchmaker have done—not a blind one, but one with keen eyes and an excellent loupe—if called upon to design a microbe that would thrive among people so fortified by vaccines and antibiotics that, to judge by their actions, they had lost all fear of germs? Nature got there without the loupe. It cooked up an all-purpose anti-vaccine, so tiny and gentle that it spread unnoticed for decades, and so innocuous that it never quite gets around to killing you. It leaves that to the old guard—the bacteria, protozoa, and viruses that invade when your immune system shuts down, and feast on your brain, lungs, blood, liver, heart, bone marrow, guts, skin, and eyes.

On June 5, 1981, the Centers for Disease Control reported five cases, two of them fatal, of a rare form of fungal pneumonia that had struck "previously healthy young men" living in Los Angeles. A second report, issued a month later, noted additional clusters of Kaposi's sarcoma—a rare, aggressive form of skin cancer that is caused by a herpes virus. "The fact that these patients were all homosexuals suggests an association between some aspect of homosexual life-style or disease acquired through sexual contact and Pneumocystis pneumonia in this population," the first report declared.

These "previously healthy" patients had in fact been mortally ill for years. Isolating the underlying cause took another three years, by which time the stealth epidemic had been creeping its way across America for almost two decades. HIV has been found in the remains of a resident of Kinshasa who died in 1959. By 1985, when Rock Hudson's death gave the virus the gloss of Hollywood, twelve thousand Americans were already dead or dying. By 1992, HIV was the leading killer of young American men.

HIV isn't like cholera or smallpox. Its contagious properties aside, HIV is like cancer. It briefly causes some flulike symptoms soon after it enters the body, then typically hides unnoticed for five to ten years. It discriminates fiercely—based on ancestry, sexual lifestyle, needle-sharing lifestyle, or disability (it killed many hemophiliacs), and against babies born of the wrong mother. It typifies the diseases of the future: slow, subtle, complex, and rooted in lifestyles and genes.

The frantic demands for an antibiotic-like cure that followed the discovery of HIV reflected the old magic-bullet medical mind-set, the notion that every disease can be beaten by a molecule precisely designed to hit just the right target. HIV, we now know, doesn't fear any single drug—none has yet proved magical enough to destroy the virus for long, because HIV invariably mutates into a form that can dodge it. But while the virus was spreading unnoticed, biochemists had been developing the technologies of intelligent design. As the National Academy of Sciences would observe in 2000, the extraordinarily fast development of drugs that ended up in the cocktails that are now used to control HIV had a "revolutionary effect on modern drug design."

Those drugs arrived, however, only after HIV temporarily revolutionized Washington. The federal government had spent the stealth decades erecting a regulatory edifice that by 1981 presented enormous scientific and economic obstacles to getting these drugs developed and licensed quickly. Washington now concocted a bunch of clever ways to dodge its own rules and unleash the full fury of our biochemists against this tiny, quiet, gentle, slow-to-kill, hateful virus. And in doing that, Washington also sketched out the framework of the scientific and economic policies that we will need to unleash the far greater power of today's molecular medicine against the flaws in human chemistry that will eventually sicken and kill almost all of us.

———

As SHERWIN NULAND observes in How We Die, AIDS struck just when "the final conquest of infectious disease seemed at last within sight." Before long, we had it on Oprah Winfrey's authority that the germs were back and after us all. "One in five—listen to me, hard to believe—one in five heterosexuals could be dead of AIDS in the next three years," she declared in February 1987. Oprah was wrong. HIV was and remains tightly linked to lifestyles shared by several comparatively small, discrete groups of Americans. The virus doesn't spread fast and indiscriminately—that's the old strategy for staying out ahead of the human immune system. It specializes instead in adapting itself to cooperative hosts and outrunning their immune systems while devouring them.

HIV isn't so much a virus as a system for spawning new viruses. It replicates fast—so fast that, though the immune system fights back, it can't keep up. Each new particle formed has, on average, one mutation in its ten thousand units of nucleic acid code. "As death approaches," Steve Jones writes in *Darwin's Ghost*, "a patient may be the home of creatures—descendants of those that infected him—as different as are humans and apes."

This process allows HIV to adapt quickly to the lifestyles of different communities. On its trek out of Africa, it found shelter in polygamous communities whose naming and coming-of-age traditions often involve the sharing of mother's milk and blood. It then split into two strains, HIV-2, which predominates among African heterosexuals, and HIV-1, which favors homosexuals. New subsidiary clusters—HIV-1A, 1B, and so forth—continue to emerge. The predominant form of the virus found in Thailand soon after its arrival in the early 1990s was "something of a specialist at travel by the anal route." A decade later, "in its new nation of sex tourists," that strain had given way to a variant that "prefers conventional sex." In varying degrees, all sexually transmitted germs do much the same. "Every continent, with its own sexual habits, has its own exquisitely adjusted set of viruses."

In the 1970s San Francisco's gay community offered HIV so many opportunities to spread that its best strategy was to attack aggressively, replicate fast, and move on, so it grew steadily more lethal. When its hosts grew more cautious, HIV spawned new strains that killed more slowly, giving the virus more time to spread.

Human genetic diversity also influences how quickly HIV kills. Some people develop AIDS quite soon after they are infected, others much more slowly, and "elite controllers" carry genes that apparently allow their immune systems to keep HIV under control indefinitely. Three genes appear to be involved; about a quarter of all Asians, and smaller fractions of Africans and Europeans, carry favorable variations. One is apparently the relict of an ancient viral infection, possibly smallpox. In 2007, doctors transplanted stem cells donated by an elite controller into the bone marrow of an HIV-positive leukemia patient. The patient stopped taking HIV drugs the same day. Four years later, doctors were unable to find any trace of the virus in his body.

The HIV pandemic will now be exerting selection pressure on humanity, favoring the genes that allow its hosts to survive long enough to bear children. Some of those hardy children will carry imprints of HIV itself in their genes. Like other retroviruses, HIV replicates by inserting a template of itself into the host cell's DNA, creating a hybrid human-virus cell that then churns out new copies of HIV in a cancerlike frenzy.

Retroviral diseases and cancers thus have something fundamental in common: both are propelled by flaws embedded in our own genes. And thus a failed cancer drug emerged as the first successful HIV drug. It had been sitting on the shelf, unlicensed, for twenty years.

IN THE 1940s, at Burroughs Wellcome, George Hitchings and Gertrude Elion, whom we met briefly in the preface to this book, had started searching
for drugs that might exploit differences between the nucleic acids of healthy
cells, cancer cells, and germs. Because cancer cells replicate very fast, Hitchings and Elion began pursuing drugs that might interfere with the enzymes
that help replicate DNA. As Elion would later recount in her Nobel Lecture, they focused on purines, a family of chemicals that are incorporated
in two of the letters of the four-letter DNA alphabet, and two letters of the
similar RNA alphabet used by certain groups of viruses—among them HIV.

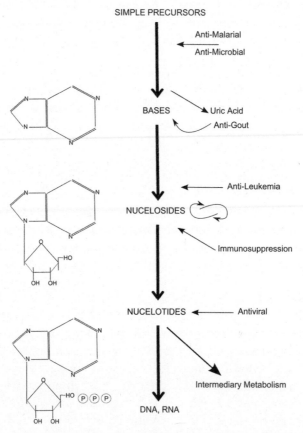

FIGURE 7.1   The purine path to chemotherapy. Purine bases are involved in a sequence of reactions that assemble DNA and RNA.
Drugs that are structurally similar to the intermediate molecules can
block the intervening steps and thus disrupt biochemical pathways
associated with a wide variety of different diseases.
Source: Nobel Assembly, "The Nobel Prize in Physiology or Medicine 1988"
(press release), www.nobelprize.org/nobel_prizes/medicine/laureates/1988
/press.html.

Their four-decade pursuit of purines would span many of the major types of pathology that surfaced after the demise of the legacy germs. The idea soon led them to two leukemia drugs. Then to a pair of drugs that attack the same enzyme, one to treat malaria, the other aimed at various bacterial infections. Then to drugs that suppress the immune system—drugs still used in the treatment of autoimmune diseases and to prevent rejection of transplanted organs.

Developed elsewhere by Jerome Horwitz, zidovudine was another early product of the approach that Hitchings and Elion had pioneered. But 1964, with memories of thalidomide still fresh, was a bad time to be pursuing drugs that tampered with reproductive chemistry, and zidovudine proved quite toxic in early animal tests. Some years later a German lab found that zidovudine curbed a viral infection in mice. Then Burroughs Wellcome acquired the drug—Hitchings and Elion had become interested in antiviral drugs when one of their leukemia drug prototypes showed promise when tested against the live smallpox vaccine. A team led by Elion explored the possibility of using zidovudine to treat herpes but set it aside in favor of a more promising alternative; in 1977, their work finally culminated in the first effective herpes drug.

AIDS surfaced four years later, and Washington was suddenly begging drug companies and researchers with virus-killing expertise to send in whatever they might have on the shelf for testing in the government's secure HIV labs. A biochemist at Burroughs sent zidovudine to scientists at the National Cancer Institute and Duke University. In lab tests, the drug looked promising.

But HIV drugs presented a delicate problem. HIV is invisible and usually harmless for the first five or so years of the infection. What if zidovudine caused grave side effects that took four years to surface? In Washington, a drug had to prove, in meticulously choreographed trials, that it would deliver clinical benefits in live patients. Zidovudine couldn't prove it was good for patients, at least not to Washington's satisfaction, any faster than HIV killed them.

Presented with a dreadful but slow-motion disease, the first antidote that showed real promise in the lab, and plausible biochemical logic for why the drug might work, the FDA scrambled to draft new protocols that would allow clinicians to dodge questions that conventional trials couldn't answer quickly—questions about long-term safety effects, among others. But many patients who had been infected with HIV years earlier weren't going to live long enough to thank the lawyers for scrambling. So the FDA approved a first zidovudine trial limited to patients suffering from fungal pneumonia, one of the most common killers when the infection turns into full-blown AIDS. The zidovudine trial had to be terminated prematurely when the dead-patient count reached nineteen to one against the placebo. Doctors can't ethically keep prescribing a placebo just to run up the score once it becomes clear, to them at least, that the drug being tested works.

A drug that had arrived at the HIV lab in February 1985 was thus licensed in March 1987, far more quickly than any comparable drug had been approved since 1962. Hitchings, Elion, and a third structural design pioneer shared the Nobel Prize for Medicine a year later. Zidovudine is better known today as AZT.

———

HALF A CENTURY into the business, Washington thus demonstrated that it had a good system for licensing a new drug that suppressed a slow virus—so long as the drug could be tested in patients who were also infected with a fungus that was about to kill them, and was then licensed for their use alone. But as the FDA knew full well, other HIV patients weren't going to wait. There immediately followed what has been described as a froth of therapeutic euphoria, and evidence that AZT worked rapidly accumulated. Three years later the FDA broadened AZT's license to cover early-stage treatment.

It had taken thirty years for humanity to discover that a dreadful new virus was quietly spreading around the globe, and another five for the United States to start fighting the virus with a twenty-five-year-old drug. HIV outwitted AZT in twenty months. That should have come as no surprise. After all, the virus had outwitted the human immune system, which has the capacity to churn out vast numbers of different magic bullets in a matter of days.

While HIV was laughing its way past AZT, biochemists were analyzing other aspects of the virus' protein chemistry. Captopril, Squibb's blood-pressure drug, had blazed a human-protease trail through Washington just a few years earlier. Other biochemists now isolated the gene for a protease enzyme that HIV uses to assemble its protein shell, manufactured the enzyme itself, worked out its three-dimensional structure, identified a key point of vulnerability, developed the first HIV-protease inhibitor, and completed a lightning-fast trip through the FDA in 1995. Other protease inhibitors soon followed.

Within months HIV was developing resistance to one of the protease drugs, too. All over the world, the virus picked up the same four mutations at different points in its genome. It had to pick them up in a particular order, and it did so, everywhere. The four mutations couldn't all have occurred in any single patient—even HIV doesn't mutate fast enough for that to have happened by chance. But the virus didn't have to finish the job inside a single patient.

A third class of HIV drugs that target another fragment of HIV's chemistry emerged while this was happening. Doctors and patients then began mixing up three-drug cocktails and zeroed in on one that reduced virus counts in the blood to undetectable levels. Today's HIV drugs target nucleosides, nucleotides, and three other classes of enzymes (transcriptase, protease, and integrase); "fusion" or "entry" inhibitors bind to glycoproteins in ways that stop the virus from prying its way into our cells.

THE RAPID ARRIVAL of these precisely targeted drugs launched the development of what would become a paradigm of how modern molecular medicine arrives at treatment regimens that are as safe and effective as the best drugs and the best medical science can make them. The HIV drugs that quickly followed AZT were also the beneficiaries of two major loopholes in the federal drug law created soon after HIV surfaced. In practice, though not framed in these terms, the loopholes got the FDA fairly close to what might be called tool kit licensing: License a drug not as an antidote to clinical symptoms, but as a molecular scalpel or suture, and let doctors take it from there. Doctors prescribe the drug to patients whose disorder presents the target that the drug is known to hit, perhaps in combination with other drugs directed at other targets. They work out the connections between molecular and clinical effects on their own, one patient at a time. The FDA-approved label plays little if any role.

Enacted in 1983, the Orphan Drug Act (ODA) took a first significant step toward tool kit licensing. The original idea was to help resurrect "orphan drugs," those dropped by pharmaceutical companies because too few patients needed them, but the law ended up covering other drugs that addressed rare and currently untreatable diseases. It directed the FDA to be very flexible when licensing such drugs—orphan drugs may be licensed on the strength of favorable case reports, animal models, or even in vitro studies when no good animal model exists.

While the FDA was relying on a fungus to help rush AZT through Washington, the agency was also drafting its "accelerated approval" rule, which was finalized in late 1992, codified and somewhat expanded by Congress in 1997, and endorsed again in 2012. When the disease is sufficiently serious and available treatments are inadequate, a new drug can get to market by demonstrating that it does indeed produce its intended molecular-scale effect or, more generally, produces favorable changes in what the FDA calls "biomarkers" or "surrogate end points."

Biomarkers and other surrogates allow the FDA to make a first call about the drug's efficacy much earlier, without waiting for clinical effects

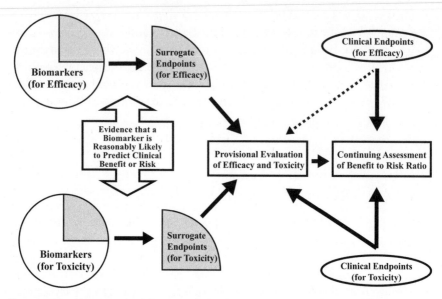

FIGURE 7.2   A conceptual framework for biomarkers and surrogate endpoints.
Source: Arthur J. Atkinson et al., "Biomarkers and Surrogate Endpoints: Preferred Definitions and Conceptual Framework," *Clinical Pharmacology and Therapeutics* 69 (March 2001): 89–95.

to surface and persist for some (arbitrary) period of time. In the Gleevec-versus-leukemia trials launched in 2000, for example, doctors tracked the drug's performance by following blood counts (hematologic response) and the number of cells bearing the Philadelphia chromosome (cytogenetic response). The FDA must be persuaded that the encouraging microscopic change is connected to the clinical symptoms, but "reasonably likely" will suffice. The truncated front-end trials need not resolve concerns about how the drug's performance might be affected by many aspects of genetic or life-style diversity; "differences in response among subsets of the population," in FDA parlance, may be addressed later. So, too, may open-ended questions about distant shadows of side effects. The manufacturer must still complete controlled trials, but does so after the drug is licensed—and thus does so after, or in tandem with, the wider use of the drug by unblinded physicians who can investigate why the drug works in some patients and not others as they treat. The FDA rescinds the license if things don't pan out.

Almost all of the early HIV- and AIDS-related drugs were designated as orphans. Most were rushed through the FDA under the accelerated approval rule. And almost all were widely prescribed off label.

HIV'S ENDLESS MUTABILITY, and the arrival of precisely targeted drugs that were licensed quickly, largely on the strength of their biochemical logic and low-level effects, left doctors in charge of working out the molecular details that determine how to combine drugs into the cocktail that produces the best clinical outcome for the individual patient. We have since learned how that process unfolded and where it ends. It has unfolded remarkably well. And it doesn't end.

A couple of dozen HIV drugs have been approved worldwide; they are typically used in about ten fairly standard cocktails. The efficacy of each cocktail depends on which strain launched the infection and how it has evolved inside the patient. Different forms of the disease predominate in different countries, and also track gender, sexual practices, needle use habits, and other factors, including the variations in human genes that give different individuals more or less inborn resistance to the infection.

When HIV is viewed from the treatment perspective, we see that medicine is now dealing with at least ten different diseases, each forever poised to mutate into some new, untreatable form. Treatments work best when the doctor selects just the right trio of scalpels from the molecular tool kit. Selecting them isn't easy, because so many different variables come into play. Until recently, trial and error played a large role. The doctor started with one mix, monitored viral loads and other biomarkers in the patient's bloodstream, and adjusted the treatment accordingly.

Monitoring and adjusting on the fly remain essential, but the process is now often guided by extremely sophisticated analytical engines fueled by huge collections of patient records that include data on HIV genotypes, treatment histories, and responses, along with patient age, gender, race, and route of infection entry; patient genotypes are certain to be added soon. As of late 2011, the largest such engine—Europe's EuResist network—was using data from 49,000 patients involving 130,000 treatment regimens associated with 1.2 million records of viral genetic sequences, viral loads, and white blood cell counts.* As described by its manager, the network is "continuously updated with new data in order to improve the accuracy of the prediction system." When presented with twenty-five actual case histories that weren't already in its database, EuResist beat nine out of ten international experts in predicting how well the treatments had performed. The study was dubbed "Engine Versus Experts."

---

* "The EuResist system is a linear combination of three statistical learning engines predicting response to treatment. . . . The models combined in the final prediction system include: a Generative Discriminative engine using a Bayesian Network prediction as an additional covariate, a Mixed Effects engine modeling interactions between covariates, and an Evolutionary engine exploiting the genetic barrier to drug resistance as an additional predictor."

Here, then, we have a medical world that stands Washington's regulatory science on its head. Whatever we may call it up here, there is in fact no single disease down there, and the disease down there tomorrow will be different from today's. We have treatments that work, but no single HIV drug can honestly be called "safe" or "effective." All have nasty side effects, and when they are used one at a time, each may fail the individual patient—and endanger others by helping to breed a new drug-resistant strain of the virus.

According to one recent count, the FDA has also licensed or at least tentatively approved eight cocktails. Controlled clinical trials were completed before most of the later-developed drugs were licensed, and once it was clear that only cocktails worked, new drugs had to be tested as part of cocktails from the get-go. But nobody even bothered to pretend that when the FDA licensed them, it had in hand what the federal drug law requires—"substantial evidence" about how these cocktails would perform when they met any substantial fraction of all the variations in HIV and patient chemistry that they might encounter in the future.

The virus continues to evolve, so the cocktails will remain "safe" and "effective" in any meaningful sense of those words only so long as we continue to prescribe them as directed by continuously updated databases. HIV can always gain time by killing its host more slowly—that gives it more time to evolve in today's host, and also more time for that host to infect his or her successor.

ONCOLOGY WAS THE other field in which targeted drugs and regulatory loopholes quickly started playing a large role. For a time, many cancer drugs also got the benefit of the Orphan Drug Act and the FDA's accelerated approval rule, and as a result, many drugs, after a first round of screening for safety issues, were licensed largely on the strength of early indications of efficacy in even a minority of the patients treated. Oncologists also routinely take advantage of the older and much broader loophole that allows doctors to prescribe licensed drugs off label.

To begin with, molecular medicine was able to transform older drugs—estrogen, for example—into precisely targeted tools by working out precisely how they interacted with cancer cells down at the molecular level. And as we have seen, drugs such as Gleevec and Herceptin provided dramatic early demonstrations of the power of structure-based design and monoclonal antibodies. The power of those technologies has improved rapidly ever since. The 2007 discovery of a gene associated with one form of non-small-cell lung cancer led to successful clinical trials of Xalkori three years later and the FDA's accelerated approval of the drug in 2011. "Once we understand a cancer cell, we can come up with a treatment very quickly," declared Dr. Mark Kris, a lung cancer specialist at Memorial Sloan-Kettering Cancer Center in New York.

Here again, the arrival of precisely targeted drugs, their flexible, accelerated passage through the FDA, and the off-label loophole left it largely to oncologists to work out, one patient at a time, how best to use them— which they do by fitting the molecular logic of the targeted drug to the biochemistry presented by the patient's cancer. Genetic matching is becoming as routine as matching a blood transfusion to blood type. At Massachusetts General Hospital, oncologists have begun sequencing the complete genome of different parts of the tumor, to launch a systematic search for multiple targets to guide treatment. Sequencing tools, together with tools for spotting the best targets for drugs to attack, are now fast, powerful, and cheap enough to make this feasible, and they are getting better and cheaper by the day.

What oncologists currently lack are enough different drugs to fire at enough of the right targets. Tumor sequencing has recently revealed, for example, that fifty to one hundred different molecular flaws can make a colon cell cancerous—far more than the eight to fifteen previously suspected. Many of these flaws, however, seem to cause trouble only when paired with one of two other mutant genes, which apparently don't cause cancer on their own. These facts point to various promising treatment strategies, but few of them can yet be tested. The search for fully effective cancer cures reminds one researcher of the early days of the fight against HIV, when AZT was the only drug on the shelf.

Working with the drugs they do have, oncologists routinely prescribe cancer drugs and cocktails far outside the boundaries that were tested in blinded licensing trials and are set out in the FDA-approved label. A nonprofit alliance of twenty-one leading cancer centers evaluates and publishes information on off-label uses. Off-label and cocktail therapies sometimes end up being steered through the rigid, slow, and expensive trials scripted by the FDA. But as a practical matter, the vast majority never will be— cancer cells, like HIV, have limitless power to reconfigure their chemistry, they do so in ways that vary too much from patient to patient, and there are therefore too many combinations of drugs, dosages, and patient profiles to explore and calibrate.

In early 2013, IBM announced the arrival of a new engine—"Interactive Care Insights for Oncology, powered by Watson"—that apparently aims to do for oncology what EuResist does for HIV. Developed in partnership with WellPoint and Memorial Sloan-Kettering, and powered by the supercomputer that won the engine-versus-experts challenge on *Jeopardy*, the engine was initially drawing on "600,000 pieces of medical evidence, two million pages of text from 42 medical journals and clinical trials in the area of on-cology research. Watson has the power to sift through 1.5 million patient records representing decades of cancer treatment history, such as medical records and patient outcomes. . . . Watson continues to learn while on the job, much like a medical resident, while working with the WellPoint nurses who originally conducted its training."

———

THE ORPHAN DRUG Act and the accelerated approval rule have now been on the books for more than two decades. By early 1998 the FDA had granted accelerated approval to some twenty-seven cancer and HIV drugs, and sixteen drugs for other conditions, most related to one of those two disorders. In a September 2012 report, President Obama's Council of Advisors on Science and Technology (PCAST) concluded that the FDA's accelerated approval rule has "allowed for the development of pioneering and lifesaving HIV/AIDS and cancer drugs over the past two decades," and it recommended that the FDA make "full use" of accelerated approval "for all drugs meeting . . . an unmet medical need for a serious or life threatening illness."

At the FDA, however, the recent trend had been in the opposite direction, back toward conventional clinical trials. Critics of accelerated approval have focused principally on the follow-up conventional trials that the rule itself requires after the license is issued. They aren't always completed as required, and when they are, they sometimes fail to persuade the FDA that the drug should have been licensed in the first place—the promising molecular effects observed earlier don't always lead to the clinical benefits that ultimately matter.

A first, short rejoinder is the one set out in the PCAST report: while there is "some risk" in using predictive biomarkers to accelerate drug approvals, it is justified by "the opportunities for progress against serious or life-threatening diseases." While it is impossible to quantify how many lives have been saved or extended by accelerating patient access to HIV and cancer drugs, most of those drugs have since proved their worth in follow-up trials and their widespread acceptance by practicing specialists. Between 1992 and 2010, the rule accelerated patient access to thirty-five cancer drugs used in forty-seven new treatments. Twenty-six of those treatments had gone on to complete conventional follow-up trials, trials that had required a median time of almost four more years of investigation. There is little doubt that the rule has spurred rapid and important innovation in the treatment of cancer and HIV and accelerated access to drugs that have, collectively, done far more good than harm.

More fundamentally, we should now be using our experience with accelerated approval to take a closer look at the FDA's standard licensing protocols. Those protocols, as the PCAST report notes, don't allow the participating doctors to systematically explore the many molecular factors that may determine why a drug performs well in some patients and not others. Most of those factors therefore end up being explored and understood after doctors prescribe the licensed drug and researchers and companies such as IBM assemble and analyze data from these sources. But those factors play an essential role in prescribing drugs safely and effectively, so many good drugs won't make it through the licensing process *until* enough of the relevant molecular factors have been identified.

The accelerated approval framework launches the search for such factors earlier, by issuing licenses on the strength of molecular and other low-level indications of efficacy or, alternatively, clinical benefits in a minority of patients, and the license then gives many unblinded researchers and doctors the opportunity to start developing the rest of the drug-biomarker science much sooner. As discussed further in Chapters 10 and 11, the PCAST report also recommends adoption of new FDA trial protocols that would permit more systematic searches for biomarkers from the outset. "Clinical investigational studies with small numbers of patients but extensive data gathering," the report notes, can play an "extremely valuable" role in the development of drug biomarker science—as they did, for example, in the Stanford iPOP study discussed in Chapter 2.

We will return to the details in later chapters, but in brief, the FDA should—as the PCAST report concludes—be overseeing a process that allows other drugs aimed at other complex diseases to follow the trail that HIV drugs and, to a lesser extent, cancer drugs have already blazed. It is indeed true that what matters to the patient is a treatment's aggregate clinical effect on his or her entire body. But drugs deliver those effects, good or bad, by interacting with molecules down at the bottom, and that is the science on which all the rest hinges.

8

DRUG SCIENCE
FROM THE BOTTOM UP

WHEN HIV FIRST surfaced, there were no licensed treatments for the viral infection itself, nor for fungal pneumonia, Kaposi's sarcoma, or many other obscure diseases on the long list of afflictions that assailed patients when their immune systems collapsed. Many Americans who are alive today owe their survival to the audacious, articulate persistence of the gay community in the 1980s and 1990s. Their response played a pivotal role in unleashing the power of molecular medicine. When Washington got on board by writing its new rules—anti-rules, actually, that spelled out when big chunks of the old rules wouldn't be fully enforced—big drug companies delivered a slew of powerful new drugs, many of them designed from scratch using the new tools of intelligent design.

Alongside the biochemists, the most important if least celebrated players in the AIDS saga were the front-line doctors, the men and women who—endangering their own lives as they handled sharp instruments while treating their infectious patients—were determined to save lives, one life at a time, and Washington be damned. The future of medicine now hinges on our willingness to give doctors much more flexibility and responsibility of that kind going forward.

NOBODY KNOWS EXACTLY when an American doctor first treated an AIDS patient, but it was well before anyone in the world had given the syndrome a name or knew what caused it. The 1981 report that prompted Washington to point a first official finger at the unnamed epidemic described five patients who had contracted the rare form of fungal pneumonia that infects about three out of every five AIDS patients. No drug had been licensed to treat that pneumonia. The first of the five patients, however, had been treated with pentamidine, which had been developed forty years earlier to treat Gambian sleeping sickness. In the early 1970s the Centers for Disease Control, which dispenses drugs licensed in other countries for use when a tropical disease lands in a U.S. hospital, had started supplying pentamidine to treat fungal pneumonia as well. The CDC had apparently done so on the strength of a handful of reports from doctors who had used pentamidine to treat organ-transplant patients whose immune systems had been deliberately suppressed by drugs. The doctor treating the AIDS patient probably obtained it through the CDC.

Then there was the leprosy drug. It hadn't been licensed by the FDA, either, but about two hundred new cases of the disease are reported annually in the United States, and the Public Health Service (PHS) began making the drug available to U.S. doctors to treat leprosy at some point in the mid-1970s. Soon after AIDS surfaced, patients concluded that they needed that drug, too, and, unable to get it from the PHS, they began organizing buyers' clubs to smuggle it in from Brazil. Nobody knows how many patients used it to treat oral and genital canker sores, wasting syndrome, or Kaposi's sarcoma, but it turned out to be good for all three of these AIDS-related disorders. When Washington threatened to prosecute, the clubs brazenly declared that they would keep doing business with Brazil until Washington offered them a better deal. Washington relented, quietly reaffirming an existing FDA policy of not prosecuting patients who smuggled in foreign drugs for personal use. Not long after, the PHS began making the leprosy drug available to U.S. doctors to treat AIDS-related conditions. Neither pentamidine nor the leprosy drug had been put through the wringer of standard, FDA-scripted clinical trials. But two months after it licensed AZT, the FDA transformed what had been an ad hoc process of dodging its own rules one drug at a time into the first major anti-rule of the new regime.*

Under normal FDA procedures, a drug that is undergoing clinical trials is granted an "investigational" license that authorizes a tightly controlled form of treatment to a limited number of patients under blinded protocols, with half the patients typically being treated with a placebo. But the FDA has broad discretion not to enforce its rules for "compassionate" or other reasons, or to authorize most anyone—other federal agencies (the CDC and PHS among them), hospitals, even individual doctors—to "investigate" unlicensed drugs far outside the bounds set by the FDA's standard trial

---

*The FDA had done it before. In the 1970s, it had allowed the National Cancer Institute to distribute certain as yet unlicensed cancer and heart disease drugs.

protocols.* The FDA had relied on this authority to allow the CDC to start distributing pentamidine to treat fungal pneumonia years earlier. So the FDA wrote a new rule setting out when it would allow other agencies to use as yet unlicensed drugs for flexible, ostensibly investigational treatment— without blindfolds or placebos.

Collectively, the various loopholes and policies that the FDA created or expanded allowed doctors to start treating patients with some drugs (typically drugs licensed in other countries) before they had even entered the U.S. licensing process, with others well before FDA-approved trials had assessed anything much beyond short-term safety issues, and still others well before the drug company had completed the last phase of the FDA's standard, three-phase testing script. The FDA would start allowing off-script investigation "just as soon as we have the information to make a reasonable judgment." For patients at the brink (those with "immediately life threatening conditions"), that would typically mean as soon as the initial, short-term safety testing had been completed and "some evidence of therapeutic benefit" had been obtained. Patients in merely "serious" trouble would have to wait somewhat longer, but certainly not for a signature on a final license. The objective was to let doctors start to treat "as many patients as possible," as early as possible, with every "promising" drug available. The FDA conceded that the treatment-investigation was "not primarily to gain information . . . but to treat certain seriously ill patients." But it would end up doing both.

———————

IT ISN'T CLEAR what first inspired some doctor to use pentamidine, the sleeping sickness drug, to treat fungal pneumonia in the early 1970s. At that time, drug companies routinely tested many drugs developed for other purposes for possible antimicrobial effects, and doctors often did the same when trying to beat an otherwise untreatable infectious disease. This testing was frequently a process of trial and error, because biochemists and doctors usually lacked the tools to work out the biochemical rationales for fitting drugs to diseases.

Washington now demonstrated its willingness to proceed in much the same ad hoc way, informally approving use of a new drug on the strength of

———————

*This cluster of overlapping and complementary rules have been given many different names, among them "personal use import exemption," "parallel track," "open label protocols," "protocol exemptions," "open label extensions," and "compassionate use exemptions," buttressed by "fast track," "expedited review," and "priority review."

limited evidence that ordinarily it would have rejected out of hand. In 1984, for example, when it appeared that the one British supplier of pentamidine wouldn't be able to keep up with demand, the CDC simply announced it would distribute a different pentamidine salt instead—which the agency trusted because it had talked with doctors in France and Canada who had prescribed it successfully to thirteen AIDS patients. Just in time, two small U.S. companies agreed to manufacture and package the original pentamidine, and FDA licensed it in injectable form a few months later, more than a decade after the CDC had started distributing the drug in the United States.

Alongside the CDC, Washington's National Institute of Allergy and Infectious Diseases (NIAID) emerged as the main designated dodger of the old licensing regime. In the late 1980s NIAID began funding "community-based AIDS research"—studies of not-yet-licensed drugs in doctors' offices, clinics, community hospitals, drug addiction treatment centers, and other primary care settings. One objective was to "offer greater treatment access for groups of AIDS patients that have not always had full opportunity to participate in existing studies: intravenous drug users, blacks, Hispanics, women (including pregnant women, whose babies are at risk of being born with AIDS), and those not living near major research facilities." A second objective was to catch up with the medical scofflaws. Treat-and-learn programs could involve both drugs that had proceeded through some of the early stages of FDA licensing process and any "drugs or therapies currently in wide use—whether they've been formally studied or not."

The first treat-and-learn drug that NIAID began distributing was Neutrexin, for use in treating fungal pneumonia in AIDS patients who couldn't tolerate pentamidine. The FDA would license the drug five years later. A NIAID-sponsored consortium of three hundred San Francisco doctors established that by using a nebulizer to administer a monthly puff straight to the lungs pentamidine worked prophylactically, and the FDA expanded the drug's license to cover that mode of delivery, too. In doing so, the FDA acknowledged that the long-term risks of inhaling the drug were unknown, and probably would never be investigated the old way, either. The San Francisco crowd had learned too much, the word had spread, and neither HIV-positive patients nor their doctors would be willing to participate in standard, double-blind trials.

By 1995, the FDA had granted treat-and-learn licenses for twenty-nine drugs, twenty-four of which went on to complete the FDA licensing process successfully.

NOW, BACK TO the leprosy drug. Here, the process of learning while treating followed a somewhat different but even more instructive trajectory. It would end up bridging the old world of perilous guesswork, with its sometimes lucky but sometimes tragic results, and the new world of biochemical logic.

In 1964, Jacob Sheskin, an Israeli physician, had just admitted to his ward a frantic woman suffering from the excruciatingly painful eruptions that often develop in the later stages of leprosy. In an attempt to calm her down, he prescribed a sedative that he happened to find left over on his shelf. Overnight, to his astonishment, her skin lesions and mouth ulcers were dramatically reduced. Initially, Dr. Sheskin's colleagues didn't believe his claim—they couldn't imagine how a sedative could help treat a bacterial infection. But Brazil, which had the second-highest rate of leprosy in the world, began using the drug widely in 1965, and Dr. Sheskin conducted successful clinical trials in Venezuela, where leprosy was also common. Medical science, however, still had no clue why the drug worked, nor did it have the tools to find out.

By the late 1980s, it did. The drug didn't attack the leprosy bacterium; it alleviated symptoms that develop when the infection sends the human immune system into overdrive. Dr. Gilla Kaplan, an immunologist at the Rockefeller University in New York, tracked the connection to tumor necrosis factor (TNF), one of three intercellular signaling proteins (cytokines) that the drug suppresses. TNF, it turns out, plays important roles in the communication system that the body uses to fight both germs and cancerous human cells. But when engaged in a losing battle, the body sometimes produces too much TNF, which can then cause painful lumps and lesions on the skin. TNF overloads can also cause wasting syndrome, a common condition in the late stages of AIDS.

Other hints that the leprosy drug was calming down the immune system were emerging at the same time, and AIDS doctors, who knew that their patients often developed autoimmune diseases, quickly grasped the implications and began prescribing the drug to treat AIDS-related disorders. In 1990, French doctors reported that the drug had proved effective in treating painful oral and genital canker sores in seventy-three AIDS patients. In 1994, several teams reported that the drug had reversed wasting syndrome. Other doctors began looking for additional TNF-related problems and were soon investigating the drug's effects on various skin disorders and other inflammatory conditions, as well as autoimmune diseases such as lupus and rheumatoid arthritis.

As noted earlier, Washington, through the Public Health Service, had begun making the drug available to U.S. doctors not long after other countries began using it to treat leprosy. The PHS then extended that policy to AIDS-related conditions soon after doctors first reported success in prescribing the drug to treat them. NIAID quickly joined the fray, sponsoring various studies and trials to explore the drug's molecular and clinical effects, among them some conventional, placebo-controlled trials. In 1995, the FDA, in an action that set the stage for the most brazen dodging of its own rules yet, had asked several companies to consider cashing in on the leprosy epidemic that wasn't sweeping across America. Celgene responded and conducted a seventeen-patient study in the Philippines, and the drug—designated an orphan by the FDA—was quickly licensed for sale in the United States in 1998. But only to treat leprosy. As everyone knew would happen, sales boomed anyway, overwhelmingly to AIDS patients who used it off label.

Meanwhile, other doctors had discovered that the leprosy drug inhibits the development of healthy new blood vessels. To starve cancerous tumors, oncologists began adding the drug to some of their cocktails. The first reports of good results were published in 1999.

AIDS patients, as we have seen, are particularly susceptible to Kaposi's sarcoma, a previously rare form of skin cancer. An unusual cluster of eight cases reported in early 1981, shortly after the first fungal pneumonia report, had played a key role in exposing HIV. A medical report describing successful use of the leprosy drug in treating Kaposi's was published soon after Celgene secured the leprosy license, and until overtaken by other drugs, the leprosy drug played an important role in treating Kaposi's sarcoma in AIDS patients. Having jolted the FDA out of its thalidomide-induced paralysis, HIV thus helped resurrect thalidomide itself—the leftover sedative that Dr. Sheskin had pulled off the shelf forty years earlier.

Celgene has developed a structurally different (optical isomer) variation of the thalidomide molecule that is much more potent but doesn't cause birth defects—at least not in lab tests. In 2005, the FDA licensed it to treat a bone marrow disorder that often causes severe anemia and leads to one form of leukemia in about a third of those patients. A 2006 license covers use in treating another blood and bone marrow cancer that kills more than ten thousand Americans a year. Before the license was issued, thalidomide's annual sales—still, officially, for use as a leprosy drug—had risen to $300 million.

————

AS SOME CRITICS saw it—and still see it to this day—the learn-while-you-treat scramble was a reckless retreat to the unscientific past. It was indeed a retreat, but far less reckless than it would have been a decade earlier—because it was in fact a retreat to the scientific future. When left entirely to their own devices, AIDS patients did indeed pursue all sorts of junk therapies: laetrile, derived from apricot pits, was one notorious example. But by the time AIDS surfaced in 1981, front-line doctors could start prescribing drugs in much the same way as biochemists were now designing them, guided by the molecular logic of the disease. And when doctors and biochemists joined forces, the track record in developing drug science from the bottom up was remarkably good.

Guided by the extraordinarily sensitive and precise molecule-tracking tools that have emerged since Dr. Kaplan linked thalidomide to TNF, countless other doctors now engage daily in the adaptive development of drug science, one patient at a time. The only way to start treating an officially untreatable disease—which is to say, any disease not addressed on an FDA-approved label—is to search for a drug that makes molecular sense and, if one can be found, prescribe it off label. With their patients' consent, that is what many excellent doctors often do. The results often then trickle into the medical literature as case reports. Many are soon forgotten; some start a process that culminates in important new treatments, widely used.

The patient-by-patient scrutiny of both molecules and clinical effects is the process that, over time, will allow medicine to zero in all the molecular targets that define each distinct form of a disease along with all the other biomarkers that may affect each drug's performance. When things work out well, other doctors hear about it and follow. And when their collective experience becomes a reasonably firm medical consensus, Washington has a problem. Patients who are convinced of a drug's efficacy won't sign up for a coin toss on a sugar pill. Doctors who are equally convinced can't ethically toss that coin just to humor some statistician in Washington. Washington has some power to impede off-label use of a licensed drug, but that is often politically impossible once doctors and patients find out and come to rely on that off-label use.

Thalidomide's effects on developing fetuses could not have been discovered by the testing required under the 1962 drug law that the drug itself launched. For ethical reasons, pregnant women must be excluded from almost all formal clinical trials, and drug licenses are narrowed accordingly. So when pregnant women need drugs, off-label use is often the only option available. The individual doctor and patient must then decide whether to conduct the off-label trial that may keep the pregnant mother with leukemia alive long enough to deliver her child, or perhaps keep the HIV-positive mother alive and also protect the child from getting infected while still in the womb. Much of the time drugs can't ethically be tested in infants, either. In 1998, doctors struggling to save premature babies, who often have trouble breathing, remembered that a drug widely used by adult men had originally been developed to help blood flow to oxygen-starved heart muscles. The drug in question—Viagra—hadn't been licensed for that purpose, but with the lives of their tiny patients at stake, doctors tried it anyway. It seems to work quite well, and has since been licensed for that use in adults.

————

AIDS was so dreadful, its victims so young, the HIV epidemic so frightening, that the ethics and potential harms of donning blindfolds and plodding through FDA-scripted trials simply could not be ignored. AIDS patients and their doctors thus pioneered the modern process of developing drug science from the bottom up—and did so well before blogs, Twitter, and computers powerful enough to run the EuResist engine arrived to help. They exploited every dodge that would let them develop the drug science on their own, far out ahead of Washington's crowd doctors. However fast Washington moved, however much it lowered the regulatory bar, patients and doctors moved faster. Some patients were dismayed when Washington began legalizing the use of unlicensed thalidomide and—in an attempt to prevent its use by pregnant women—began closely tracking who was using it for what purposes. "We are concerned that removal of thalidomide from the buyers' clubs will make it more difficult to explore new possible uses for the drug," one observer wrote in 1995. "Official clinical research is usually years behind the leading front-line scientists, physicians, and patients."

But by then the front line was largely running the show. Drug companies quickly became engaged once Washington made clear how warmly it would welcome them. New treatments for HIV and AIDS-related diseases then emerged remarkably fast—fast not only by Washington's standards but by medicine's as well. By many measures, molecular medicine advanced more rapidly during this interlude than it had at any previous time in history.

Other parts of Washington were persuaded that we needed more of the same. One provision of the FDA Modernization Act of 1997—passed without a single dissenting vote in Congress, with Senator Edward Kennedy a key player in forging the consensus—codified or implicitly endorsed the rules the FDA had started drafting soon after HIV swaggered into town. The new law specifically expanded the scope of the accelerated approval rule, which would now apply not only to conditions the FDA judged to be "serious" or "life-threatening" but also to disorders likely to get progressively worse or end up impairing day-to-day functioning—a standard that can be read to include Alzheimer's, coronary heart disease, and many other diseases.

But the 1997 act neglected to make the modernization mandatory. Had it done so in strong language, the FDA might have set about systematizing the learn-as-you-go process, and molecular medicine would now have in place comprehensive data pooling and analytical engines to do for much of the rest of medicine what the EuResist engine does for the treatment of HIV infections. Molecular medicine would be advancing much faster as a result.

Instead, the rules developed to accelerate the battles against HIV and AIDS receded as effective HIV treatments were developed. The FDA quietly curtailed the parts of those rules that allowed unblinded doctors and patients to start exploring the molecular interface between drug and patient much earlier. In 2009, the FDA amended the rule on investigational treatment it had promulgated almost two decades earlier. In its current form, the rule allows broader distribution of a drug only when conventional, blinded trials are already quite far advanced, and only when the FDA is convinced that the broader distribution won't "interfere" with the completion of the trials already under way or any that might be initiated in the future to expand the drug's license. Similarly, when the FDA continued to apply the accelerated approval rule, it usually did so only on the basis of fairly advanced clinical surrogates—the shrinkage of tumors, for example—rather than molecular or cellular surrogates.

The FDA's reactionary inclinations seem to have grown stronger, rather than weaker, as the molecular science has advanced. It is not difficult to understand why. Medicine's ability to generate torrents of data that are relevant to a drug's performance is expanding far faster than Washington's ability to make sense of it all. And the gulf will only widen from here on out.

We can start creating the databases that will eventually tell us when a drug—newly developed or used in a new way—can perform safely and effectively when the first patient is treated anywhere in the world, either on label or off. The databases can incorporate increasingly detailed biochemical profiles of patients and drug regimens, together with medical outcomes, and the data can be updated continuously. By analyzing such databases, biochemists and doctors expose key molecular patterns that affect the drug's interaction with the disease and bystander chemistry in the rest of the patient's body, and progressively refine how the drug is prescribed going forward. As experience accumulates, medicine—guided by digital engines—will grow increasingly good at prescribing one or more drugs in ways that produce predictable effects in future patients. But the complexity and variability of molecular medicine rises mind-numbingly fast when it tackles biochemically complex diseases and side effects and takes into account changes in the patient's body caused by diet, smoking, and numerous other factors that can affect the biochemical environment in which the drug operates.

By retreating into the dark comfort of blinded trials that track clinical symptoms, the FDA solves the complexity problem by refusing to get mired in molecular details—the details on which the progress of the disease and the efficacy and safety of targeted drugs actually hinge. By late 2010, the FDA's thalidomide brain was dominant once again, and for the most part, the agency's old testing protocols were being enforced as rigidly as ever. Minor deviations in scripted protocols were being cited as reasons for rejecting very promising results, some involving drugs for deadly diseases for which there is no other good treatment. In 2010, for example, the FDA rejected the only medication able to treat a rare, fatal lung disease (idiopathic pulmonary fibrosis) because the drug—which had already been licensed in Japan and was about to be licensed in Europe—had completed only one of the two large studies required by the FDA. The fraction of cancer drugs granted accelerated approval had been cut in half. Legislation enacted in 2012 attempts yet again to nudge the FDA back to the future—but once more leaves it all to the agency's discretion.

9

THE FADING MYTH OF
THE FDA'S "GOLD STANDARD"

THE FDA'S FRONT-END, double-blind clinical trial protocols, it is often said, establish the "gold standard" for drug science. Washington's meticulous scripting of those protocols, however, was launched in 1962 by a drug that vividly illustrates why it is so often impossible to script them well. When the cause-and-effect connections are complex, writing a good trial script requires information that only the trial itself can reveal.

Thalidomide did no significant harm to most adult women (though there were early reports, since confirmed, of adverse peripheral nervous system effects in some patients), and it harmed their unborn children in many different ways, but only in the early weeks of pregnancy. Statisticians call it the "reference class problem," or the problem of "external validity." We don't know when a drug's performance in the group of human bodies assembled for the launch of today's clinical trial will provide a good prediction of how the drug will perform in future patients until we know how many patient-side biochemical factors can affect the drug's performance and how evenly or otherwise those factors are distributed among patients who will end up using the drug.

When scripted as the current FDA protocols require, drug trials routinely make unsupported and unacknowledged assumptions about those issues, and thus quietly ignore uncertainties about the model of reality that determines how to gather data and how much to gather. The relentless growth of the FDA-mandated clinical trials since 1962 reflects the emperor's own dawning realization that his wardrobe was furnished by Victoria's Secret. Washington began losing confidence in quick, small clinical trials as science began to expose the slow, complex diversity of human chemistry. In the last decade, our newfound power to scrutinize everything down at the molecular level has exposed vastly more biochemical diversity and complexity. And any molecular difference between two bodies might be the difference that allows the same drug to perform well in one and badly in another.

———

GIVEN WHAT WE now know about the environments in which drugs operate, the unresolved question at the end of many failed clinical trials is what exactly failed: the drug or the FDA-approved script, which specifies how many patients will participate in trials, how those patients will be selected, and how long the trials will last. It's all too easy for a bad script to make a good drug look awful. For example, the script begins with a standard clinical definition of a disease that is in fact a cluster of many biochemically distinct diseases; a coalition of nine biochemical minorities, each with a slightly different form of the disease, vetoes the drug that would help the tenth. Or a biochemical majority vetoes the drug that would help a minority. Or every component of what would be an effective cocktail fails when each drug is tested alone. Or the good drug or cocktail fails because the disease's biochemistry changes quickly, but at different rates in different patients, and to remain effective, treatments have to be changed in tandem, but the clinical trial is set to continue for some fixed period that doesn't align with the dynamics of the disease in enough patients. Or side effects in a biochemical minority veto a drug or cocktail that works well for the majority.

Our bodies are packed with biochemicals that keep us healthy when they work together in just the right balance but kill us when the balance gets disrupted. Many of the cocktail cures we need may likewise comprise drugs that can be developed systematically, one by one, to control promising molecular targets but can't deliver any useful clinical effects until combined. Getting that kind of medicine through today's FDA would be, for all practical purposes, impossible. Even if it were allowed (and it probably wouldn't be), running numerous combinations of several new drugs through FDA-approved trials in the hope of finding the combinations that will fit the countless biochemical configurations of patients that they might benefit would take too long and cost too much.

The FDA itself has implicitly acknowledged this problem in the rare instances where it has approved gene therapy—using a virus to insert a gene into the patient's cells. No single gene can do any good on its own, but added to the patient's own biochemical tool kit, it can. Similarly, Provenge and some of the other promising cancer vaccines rely on a process of adding new biochemical components to the patient's own immune system cells. There is in such cases no licensable, stand-alone drug anywhere in sight, still less one that can prove its worth in a standard clinical trial. What gets licensed is a systematic process for reconfiguring certain parts of the patient's molecular chemistry, in ways specifically tailored to that one patient.

―――――――

A RECENT REPORT from the National Research Council (NRC) includes an illustration of how we currently rely on dumb luck to help drugs that target complex disorders stumble their way through the FDA's testing protocols. Until very recently, clinicians divided lung cancers into two main types: small-cell and non-small-cell. In 2003 and 2004, the FDA granted accelerated approval to two drugs, Iressa and Tarceva, on the strength of their dramatic effects in about one in ten non-small-cell patients. Over the course of the next two years the drugs were prescribed to many patients whom they didn't help, and several follow-up clinical trials seemed to indicate that the drugs didn't work after all—probably, we now know, "because the actual responders represented too small a proportion of the patients."

Meanwhile, the report continues, the molecular disassembly of lung cancer had begun its explosive advance. In 2004, researchers identified the specific genetic mutation that activates the EGFR enzyme that these two drugs inhibit. "This led to the design of much more effective clinical trials as well as reduced treatment costs and increased treatment effectiveness." By conditionally licensing a pair of one-in-ten drugs, the FDA had launched an adaptive process that finished the job.

Under current, blinded trial protocols, however, such launches often depend on luck and circular science. The original clinical trial happens to include just enough of the right patients to persuade the FDA to license the one-in-ten drug when applying the flexible standard of the accelerated approval rule. The fortuitously and just barely successful completion of the first clinical trial starts the process that may ultimately supply the information that, ideally, would have been used to select the patients to include in that first trial.

Tarceva remains on the U.S. market, but not Iressa. In early 2005 Iressa became the first cancer drug to be withdrawn after the required follow-up trials failed to confirm its worth to the FDA's satisfaction. After further trials failed to establish that Iressa extends average patient survival, and serious side effects surfaced in some patients, the manufacturer halted further testing in the United States. The drug had, however, been licensed in Europe and other countries, subject to further study on how to identify patients it can help. So it may yet return to the United States, after doctors and patients in Europe and elsewhere finish developing the biomarker science that medicine needs to prescribe Iressa more precisely.

We do already know that Iressa survival times and side effects vary widely among patients. And we have a pretty good idea why. As Bruce Johnson, a researcher at Boston's Dana-Farber Cancer Institute and one of the doctors involved in the original Iressa trials, remarked in 2005, "For us as investigators, at this point, there are at least 20 different mutations in the EGF receptors in human lung cancers, and we don't know if the same drug works as well for every mutation . . . which is why we want as many EGFR inhibitor drugs available as possible for testing." When the FDA rescinded Iressa's license, it allowed U.S. patients already benefiting from its use to continue using it. One such patient who started on Iressa in 2004, when he had been given two to three months to live, was still alive eight years later, and walking his dogs several miles daily.

———————

THE QUALITY OF science is gauged by its ability to predict. The best predictions are now anchored in mechanistic facts about what a drug will do when it encounters various specific molecular targets. Those facts alone aren't sufficient, but they are necessary—without them, many drugs that we need can't consistently perform well, and most diseases won't be cured. As the EuResist engine and breast cancer treatment algorithms illustrate, the best predictions come from a sophisticated and continuously improving mix of the rock-solid biochemical facts and empirical data—with the mix shifting steadily toward the former.

These engines still rely on empirical data—but they do so not to pass final judgment on any single drug or drug cocktail but to reveal more complex patterns that can be used to transform the core, patient-specific biochemical facts into a personalized prediction of likely clinical effects, good and bad, that targeted drugs will have in the unique biochemical environment of an individual body. Every advance in the biochemical science diminishes the need for empirical correlations by narrowing the scope of the biochemical uncertainty. As Dr. Janet Woodcock, the director of the FDA's own drug evaluation research, put it in 2004, "Biomarkers are the foundation of evidence based medicine—who should be treated, how and with what. . . . Outcomes happen to people, not populations."

For now, at least, the only way to build a complete biomarker foundation is to study how a drug performs in as many patients as it takes to present all of the biomarkers that can affect its performance in all the combinations that occur in different patients, for as long as the biochemistry may take to run its course. We don't know how many patients must be tested, or for how long, until the tests are completed, and we don't know when those tests have been completed because biological effects can span not only lifetimes but generations of microbes and people.

The FDA doesn't know, either. And it doesn't want biomarkers involved in the licensing process until it does. That, in a nutshell, is the biggest obstacle that now stands between us and the future of molecular medicine.

WHAT THE FDA doesn't know about biomarkers is what makes its trial protocols, statistics, and licensing calls far less useful and reliable than it has long maintained. When rare but nasty side effects surface long after a drug was licensed, it is because their underlying biochemistry wasn't exposed during the trials. On the flip side, countless valuable drugs have almost certainly been abandoned not because they failed to work but because medicine hadn't yet found out how to narrow a clinical trial to include just the right patients, while Washington's statisticians insisted on expanding the trial willy-nilly. According to a recent consensus report issued by a coalition of cancer experts drawn from the industry, academia, and the FDA itself, the agency still usually relies on "traditional population-based models of clinical trials . . . designed to guard against bias of selection." Such trials "may form the antithesis of personalized medicine, and accordingly, these trials expose large numbers of patients to drugs from which they may not benefit."

That much the FDA does know. So we arrive at the FDA's new mission: licensing biomarkers. The agency calls it "validating," and the biomarkers aren't manufactured by drug companies, but these details aside, we find ourselves back in 1962. Once again, the FDA is struggling to decide how to decide when a molecule is likely to affect clinical health for better or worse. The molecules at issue now are the patient-side ones that a candidate drug will interact with, in good ways or bad.

For over two decades the FDA has accepted—in principle—the use of biomarkers in drug licensing. Define the disease and select the patients for the clinical trial using molecular biomarkers, and declare at least conditional victory when the drug performs well down at the molecular level, or perhaps at the nearby viral or cellular levels. Selecting the patients that way can make it much more likely that the drug will perform well. Judging the drug's performance that way allows the licensing call to be made much sooner.

But everything then hinges on the validation of biomarkers. Until they are pinned down to Washington's satisfaction, the message to would-be developers of drugs to treat complex diseases comes fairly close to this: *Don't even bother trying. Wait until you or someone else has worked out the molecular taxonomy of the disease and has persuaded the FDA that the science is good enough to guide the selection of patients to include in the trials. If you want accelerated approval based on molecular-scale rather than clinical effects, wait even longer—until science has nailed down the links between molecular and clinical effects.*

The agency points out that linking what happens down there to what happens up here can be very tricky, and if we get the links wrong, the FDA may end up licensing diagnostic sniffers or drugs that are useless or worse. So, as we have seen, the FDA won't approve a home-test sniffer that checks whether you carry, say, certain breast cancer genes until it's persuaded that the information supplied will be not only "analytically" but also "clinically" accurate. And the FDA won't license a drug on the strength of its molecular-scale effects until it's persuaded that those effects are "reasonably likely" to translate into clinical benefits.

Just as it was fifty years ago, the FDA's main worry is selection bias. With so many biomarkers on the molecular level, one or more undoubtedly can be plucked out to explain away the patients who keep getting sicker, and eventually you're left with only the patients who got better—whether or not the drug had anything to do with it. Good molecular medicine, however, *requires* the systematic selection of patients who present the biomarkers that allow the drug to perform well and who do not present biomarkers associated with side effects.

The FDA has been drafting rules and "guidances" (in effect, rules that can't be challenged in court) on the use of biomarkers in the licensing process since the late 1980s. The agency, however, has unlimited discretion to remain dissatisfied with the quality of biomarker science. During the decades that the FDA has been thinking about how to include patient-side molecules in the licensing of molecular medicine, researchers have identified thousands of potentially important biomarkers. Over the past fifteen years, however, the FDA has approved protein biomarkers for use in licensing at an average rate of one or two a year.

Molecular medicine often determines how strongly molecules—cholesterol, for example, or a high-cholesterol gene—are linked to clinical problems by searching for statistical correlations in large databases of patient records that include both molecular and clinical data. Strong links then point to promising drug targets. And all this can be done before a clinical trial of the drug begins.

The same statistical tools can then be used to analyze links between drug-biomarker molecular combinations and clinical effects. As it is acquired, this information can then be used to refine prescription protocols in ways that affect both efficacy and safety. Such studies have identified genetic biomarkers that can tell you in advance whether you will respond well or badly to a growing number of drugs, among them anticoagulants, antidepressants, painkillers, and drugs for heart disease, high blood pressure, hepatitis, and various cancers.

But here's the catch. Most of the drug-biomarker science can't be developed before human trials begin. FDA protocols allow very little of it, if any, to be developed during the front-end licensing trials. So, most of this invaluable predictive molecular science is developed after a drug has been licensed and prescribed to many patients, many of whom, we discover, should never have used it at all.

A drug designed to target an estrogen receptor, for example, should obviously be tested only in the ER+ breast cancer patients whose tumors present that target. But if the breast cancer drug's performance also depends on how it is metabolized in the patient's liver, as tamoxifen's does, the existence of a genetic marker that identifies the patients with the right livers often won't be discovered until doctors begin exploring why different ER+ patients respond differently to the same targeted drug. A drug's selective efficacy can also depend on a wide range of other biomarkers that are hard to identify in advance. Hitting the drug's intended target may not suffice: complex diseases may respond only to multipronged attacks, in which case the selection criteria for testing today's drug ought to include the selection of other drugs needed to complete the synergistic cocktail. Which means that it may be impossible to test the drug in the right biochemical environment until complementary drugs are available—and the same may be true for each of those other drugs.

Before a trial begins, it is even more difficult to specify selection criteria for excluding patients in whom the drug will cause serious side effects. In a tacit admission of the limits of its own trial protocols, the FDA itself helped launch a nonprofit consortium of ten drug companies and academic institutions to compile a global database of genetic links to drug side effects. In 2010, the group released data that help predict when drugs are likely to cause serious harm to a patient's liver or trigger a potentially lethal allergic response. Better late than never, but detecting these links during front-end licensing trials would have been very much better.

Many biomarkers could be detected earlier, and should be. Another way to develop drug-biomarker science is to study how individual bodies interact with drugs down at the molecular level—as was done, for example, in Stanford's iPOP study. The FDA knows that, too, and recently began approving a range of what are, by Washington's standards, innovative "adaptive" trial protocols that allow that to happen. But the FDA remains slow and reluctant to approve such trials, and unwilling to accept the complex analytical tools that extract reliable scientific knowledge efficiently from extremely complex data sets.

The problem for the FDA is that robust drug-biomarker science can't be fully developed without testing drugs in a broad range of biochemically different patients and carefully studying and comparing their responses. That means removing the FDA's cherished blindfolds and replacing simple trial protocols that analyze comparatively tiny amounts of data with complex protocols that analyze torrents—not the kind of change that ever happens quickly in Washington. As discussed further in Chapter 16, the FDA, as currently structured and funded, lacks the institutional resources—and perhaps also the expertise—to keep up with the converging, synergistic power of the biochemical and digital revolutions. Bureaucratic inertia may be a factor as well—the indiscriminate testing required by the FDA's current trial protocols is familiar, and much easier to regulate. At stake, unfortunately, is the entire future of molecular medicine.

Here, at last, the FDA is forced to confront the problem that its blindfold protocols deliberately sweep under the statistical rug. At the end of the day, the way biomarker science proves itself is by demonstrating that it can be used to write complex, personalized prescriptions that work. The "validation" that the FDA seems to be waiting for is inseparable from the actual development of drugs that interact with biomarkers, directly or indirectly, in ways that produce predictable clinical effects. There is no practical substitute for developing the drug and much of the biomarker science together—the drug-biomarker interactions are what matter. The problem is circular: we need the drug and many doctors and patients to expose and validate many of the relevant biomarkers, but at present we need the biomarkers to frame the trial that will persuade the FDA to license the drug and thus allow doctors to develop the biomarker science as they treat patients without the help of blindfolds.

During the early stages of the scramble to catch up with HIV, the FDA dodged this problem by applying the orphan drug law's licensing standard and the accelerated approval rule flexibly and aggressively. HIV's ability to reconfigure itself on the fly meant that no single drug would do much good on its own—but biochemical logic and short-term experience strongly suggested that attacking HIV's nucleic acid chemistry with one drug and its protein chemistry with another made good sense. A similar approach was soon used to accelerate the approval of many cancer drugs. The drugs were being licensed largely on the strength of their ability to interact, in some promising way, with a biochemically plausible biomarker. Without putting it in quite these terms, the FDA accepted that most of the investigation of the molecular-to-clinical links—and thus the validation of the biomarkers—would emerge later, as patients began to be treated by doctors who kept their eyes open. It did, to the advantage of many patients.

––––––

THIS GETS US back to the 2011 NRC report cited earlier. The authors sought to address the need for "a 'New Taxonomy' of human diseases based on molecular biology," and how to develop that taxonomy. Yes, we need it, the report emphatically concludes, and it therefore recommends creation of a broadly accessible "Knowledge Network," which will aggregate data spanning all the molecular, clinical, and environmental factors that can affect our health. Data from the network will help researchers "propose hypotheses about the importance of various [factors and connections] that contribute to disease origin, severity, or progression, or that support the sub-classification of particular diseases into those with different molecular mechanisms, prognoses, and/or treatments, and these ideas then could be tested in an attempt to establish their validity, reproducibility, and robustness." The underlying data and the taxonomy of diseases themselves must be "continuously" updated, tested, and refined.

In sharp contrast with the FDA, the NRC would give doctors access to new biochemical and clinical data as they are acquired, and it accepts that some doctors would then prescribe drugs (off label, presumably) on the strength of evidence that Washington doesn't yet view as "proven." "Some patients and clinicians will be more comfortable than others with making decisions that are based on clinical intuition rather than proven evidence. Any physician should be able to interrogate the Knowledge Network . . . to learn whether others have had to make a similar decision, and, if so, what the consequences were"—while also being informed whether causal links between molecules and clinical symptoms have been "rigorously validated." The NRC report expects that most doctors and patients would choose to "interact" only with the "higher-value-added levels"—presumably the "validated" levels of information on the network. But the report does not attempt to explain how the validation would happen.

The explanation is, surely, implicit in the report itself. The scientific underpinnings of molecular medicine cannot emerge from a process that demands up-front, high-certainty, crowd-based, eyes-closed, clinical-level validation, one drug and one biomarker at a time. It can only emerge from an adaptive, self-correcting process of discovery that develops an understanding of the links between molecules and clinical effects as it identifies and reassembles the pieces from the bottom up, patient by patient.

"When we were writing the report, we got into this discussion about Google Maps," says the hyperqualified Dr. Susan Desmond-Hellmann, co-chair of the task force responsible for the report and also an oncologist, former president of product development at Genentech, and chancellor of the University of California at San Francisco. "What was interesting about Google Maps is how dynamic it is. If you go someplace and they put a new street . . . it is updated for the new road. You can get into it and search it and find something in real time. What if medicine worked that way and instead of a new street, it's a new biomarker, and I have access to that pace of knowledge that could benefit the patients I am caring for? . . . Low-cost, high-throughput [genetic] sequencing and the focus on biomarkers . . . enable us to do something we were rarely able to do before: have the backbone of research be human biology."

Few doctors will view the NRC's proposal or Dr. Desmond-Hellmann's vision as radical—both address the widely recognized and rapidly widening chasm that already separates the practice of molecular medicine from the drug-patient science developed and certified the Washington way. The FDA's trial protocols start from the obsolete assumption that treating doctors, working on the patient side of the divide, can't discover and then control for molecular details that affect a drug's performance. Assume blind clinicians tomorrow, and you see only the benefits of requiring blind trials today. But molecular biology has been choreographing the action all along. Now that biochemists and doctors can see it all, design drugs to hit what they see, and track their effects down at that level, of course they do just that.

WASHINGTON IS WORKING for the benefit of tomorrow's patients, not today's. Today's patients matter so little that their doctors wear blindfolds and flip coins. But the individual patient would rather get cured than serve as a data point for the benefit of the crowded future. His doctor won't see all the relevant biomarkers that might show up in future users of the drug, but she doesn't need to. If the FDA weren't hovering nearby, the doctor and patient would remove the blindfolds, strike a private balance between what they know or can quickly find out about short-term low-level effects and what they don't know about long-term clinical effects, and get on with the process of learning a great deal about the molecular structure and dynamics of the disease in one specific body.

If all the patients involved in a trial could monitor and share their experiences, they would all adjust even faster. A placebo can't possibly be good for a patient—not unless the real drug administered to the patient down the hall is worse than useless. And in that case, the patient down the hall, for the benefit of science, is being treated with a worse-than-useless drug. Much of the time, either the drug or the placebo would wind up abandoned far sooner if the blindfolds were off.

Individual doctors and patients will *always* be able to make better decisions if they have more access to more information, even anecdotal information, so long as the facts are accurate and the uncertainties are handled intelligently. Doctors and patients can now be guided by powerful statistical tools that, as we shall see, were developed to handle extremely complex forms of uncertainty and are already in wide use throughout the rest of the information economy. These tools don't provide instant answers, but they are very good at learning step by step and, in the process, quantifying how much they know and don't know. This allows them to provide more useful information much sooner than Washington does.

Doctors are now beginning to question whether the FDA's trial protocols even remain ethical. In a 2011 essay in the *New England Journal of Medicine*, Dr. Bruce Chabner, of Massachusetts General Hospital Cancer Center in Boston, notes that in the first phase of one trial, 81 percent of patients with particular mutations responded well to a new drug for metastatic melanoma. In the third phase, patients were nevertheless randomly assigned to receive either the new drug or an older one that has a 15 percent response rate. The recent drug-innovation report issued by President Obama's advisory council, cited earlier, notes that the adaptive trials it recommends "potentially allow for smaller trials with patients receiving, on average, better treatments."

By continuing to channel the development of drug science through trial protocols developed decades ago, Washington now makes it increasingly likely that the science will never get developed at all. Blind doctors treating the crowd assembled for the FDA's benefit will lose the molecular science in the crowd. The drug won't get licensed—and the disease won't get cured.

10

ADAPTIVE TRIALS

As DISCUSSED IN Chapter 9, a recent NRC report concludes that working out the etiology of complex diseases will require the analysis of "biological and other relevant clinical data derived from large and ethnically diverse populations" in a dynamic, learn-as-you-go collaboration among biochemists, clinical specialists, patients, and others. As it happens, good drug science requires much the same—a drug is just one more molecule added to the molecular ecosystem that constitutes a body. The NRC report assumes as much when it recommends that doctors be allowed to consult the proposed Knowledge Network to find out how other patients have fared when already licensed drugs are prescribed outside the FDA-approved boundaries. As that report makes clear, the objective is "precision medicine." A molecular taxonomy of disease is only the starting point that leads to precisely targeted drugs and precise prescription protocols.

The several elements of precision medicine are tightly linked. Every time we prescribe a targeted drug, whether during a licensing trial or thereafter, we simultaneously test and have the opportunity to improve our molecular understanding of the disease that it targets. Estrogen teaches us about breast cancer receptors, for example, when it accelerates the disease in some patients while slowing its development in others. We confirm that the bacterium is the cause of the disease by targeting it with an antibiotic and watching the patient recover—or we discover that the microbe has mutated into some new form when the previously effective drug fails. Every drug is also a potential cause of the diseases that we call side effects. Tamoxifen suppresses some forms of breast cancer but raises the risk of some forms of uterine cancer. Precision medicine hinges on systematic patient selection—selection based on the drug's intended target, unintended targets associated with side effects, additional drugs that may be prescribed at the same time, and other factors. The include/exclude calls will often have to be repeated on the fly as the patient's biochemistry changes (or fails to change) during the course of treatment.

A good clinical trial of a good drug will develop prescription protocols that will make it possible for future doctors to select the patients who have what it takes to make the drug perform well. The best protocols will be based on molecular markers and effects—the whole point, after all, is not to wait for clinical effects to reveal whether the drug was prescribed well.

But as we have seen, current FDA protocols treat patient selection as a problem the drug company must solve either before the clinical trial begins or, to a limited extent, in its very early phases, which currently involve very small numbers of patients. At best, this means that the drug is prescribed to many patients whom it fails to help or even harms during the trials, and to still more of the wrong patients after it's licensed, until enough postlicensing data accumulate and reveal how to prescribe the drug more precisely. At worst, drugs that are desperately needed by some patients don't get licensed because the trials test the drugs in too many of the wrong patients. Either way, testing a drug in many of the wrong patients wastes a great deal of time and money. At some point the cost of relying on a very inefficient process to try to solidify the science up front surpasses how much the drug is likely to earn years later in the market. We then have an economically incurable disease.

In the words of Dr. Raymond Woosley, president and CEO of the Critical Path Institute, a nonprofit group established to promote collaboration among drug companies, academic researchers, and the FDA, "Randomized controlled trials are out of date, and it's time to use the tools of the future."

———————

NINE MONTHS AFTER the NRC issued its report, President Obama's Council of Advisors on Science and Technology (PCAST) released a report entitled "Propelling Innovation in Drug Discovery, Development, and Evaluation." The FDA's trial protocols, the report notes, "have only a very limited ability to explore multiple factors." These protocols lead to clinical trials that are "expensive because they often must be extremely large or long to provide sufficient evidence about efficacy." The report goes on to outline a proposal for ushering the FDA into the future of molecular medicine. The proposal has five main elements.

- The FDA should use its existing accelerated approval rule as the foundation for reforming the trial protocols used for all drugs that address an unmet medical need for a serious or life threatening illness.
- The molecular science used to select targets and patients should be anchored in human rather than cell or animal data and can be developed,

in part, during the clinical trials. Before the trials begin, statistical studies of naturally occurring genetic variations can provide valuable guidance on biomarkers to target and track, and will grow increasingly useful as databases that combine genomic and clinical data grow larger. "Clinical investigational studies with small numbers of patients but extensive data gathering" are an "extremely valuable" alternative.

- The FDA should adopt "modern statistical designs" to handle the data-intensive trials and explore multiple causal factors simultaneously— including "individual patient responses to a drug, the effects of simultaneous multiple treatment interventions, and the diversity of biomarkers and disease subtypes." These multidimensional statistical designs are much more efficient than those used in connection with the FDA's conventional protocols, and the patients involved receive, on average, better treatments.

- The FDA should also "expand the scope of acceptable endpoints" used to grant accelerated approval. Specifically, the FDA should make wider use of "intermediate" endpoints—indications that a drug provides "some degree of clinical benefit to patients" even though the benefits "fall short of the desired, long[-]term meaningful clinical outcome from a treatment." The agency has granted only eleven such approvals in the past twenty years. The FDA should "signal to industry that this path for approval could be used for more types of drugs" and "specify what kinds of candidates and diseases would qualify."

- These initiatives should be complemented by more rigorous enforcement of the existing requirement for confirmatory studies that demonstrate the clinical efficacy of drugs, and the FDA should continue and possibly expand its use of reporting systems that track both efficacy and side effects in the marketplace. The FDA should also consider a process of incremental licensing that begins with accelerated approval for use of the drug only in treating "a specific subpopulation at high risk from the disease" when larger trials would take much longer or wouldn't be feasible. The license could then be broadened to authorize broader use upon the successful completion of broader trials. The FDA would "strongly discourage"—but not forbid—off-label use in the interim.

Vigorously implemented, these proposals would go a long way toward aligning FDA regulation with the drug development tools and practice of modern molecular medicine. The accelerated approval rule puts the focus on molecular-scale or other low-level effects from the start. Protocols that allow the integrated development of drug-biomarker science lead not only to smaller, more efficient trials but to the science itself—much of which can't practically be developed in any other way. And broadening the standard for accelerated approval to include successful achievement of "intermediate" endpoints is a good starting point for addressing the most fundamental issue of all: what should it take to meet the federal drug law's demand for "substantial evidence" of a drug's efficacy in the age of molecular medicine?

The PCAST report addresses that question only indirectly. It needs to be addressed head-on, if only to make clear how important it is to persuade the FDA to implement the report's recommendations quickly and wholeheartedly.

———————

To DEAL SUCCESSFULLY with biochemically complex diseases that require complex treatments that vary from patient to patient, we will have to develop treatment regimens piece by piece, each piece consisting of a drug and a solid understanding of how a cluster of biomarkers can affect that drug's performance. Demanding a front-end demonstration that each piece will deliver clinical benefits to most patients on its own will only ensure that no treatment for the disease is ever developed. An intermediate endpoint—"some degree of clinical benefit" that indicates that the drug is interacting in a promising way with a molecular factor that plays a role in propelling a complex disease—is the best we can expect from any single piece. Even that requirement may be too demanding; used on their own, the individual constituents of some of the multidrug treatments that we need may never be able to deliver any clinical benefit at all.

A drug, for example, may successfully suppress an HIV protease enzyme and thus lower viral loads, or bind successfully with an estrogen receptor and thus shrink or slow the growth of a breast cancer tumor, yet have no lasting effect on the progress of the disease because the viral particles and cancer cells mutate their way past any single-pronged attack. It should be licensed anyway. A successful attack on a biochemically nimble virus or cancer has to begin somewhere, and the place to begin is with a targeted drug that has demonstrated its ability to disrupt some molecular aspect of the disease's chemistry in a way that had some promising effect, in some patients, at some point further along in the chain reaction that propels the disease. When the first HIV protease inhibitor established that it could do that job and thus lower viral loads, it was a drug that medicine clearly wanted to have on the shelf—even though it would take several more years to develop additional drugs and assemble cocktails that could suppress the virus almost completely and for a long time.

By allowing broader use of the drug by unblinded doctors, accelerated approval based on either molecular or modest and perhaps only temporary clinical effects launches the process that allows many more doctors to work out the rest of the biomarker science using all the tools of modern pharmacology, unconstrained by the FDA's rigid trial scripts. The FDA's focus shifts from licensing drugs one by one to regulating a process that develops the integrated drug-patient science necessary to arrive at the complex prescription protocols that can beat biochemically complex diseases. As the PCAST report notes, the FDA already has, and should continue to exercise, the authority to monitor and regulate that follow-up process and to modify or rescind the initial license if the clinical benefits don't materialize or if serious side effects do.

Adaptive trials can be structured in many different ways, and the details are beyond the scope of this book. (The PCAST report includes a description of the I-SPY 1 [2002–2006] and I-SPY 2 [ongoing as of early 2013] trials of breast cancer drugs.) In brief, adaptive trials gather a great deal of data, tracking genes, proteins, microbes, and other biomarkers that control the trajectory of the disease and cause different patients to respond differently to the same treatment regimens. The protocols evolve as the trial progresses and the collective understanding of the drug-patient molecular science improves.

Data-pooling networks and pattern recognition computers should be used to systematize the process from the outset. Informed by a constantly expanding database of patient experience, the computers will be engaged in the rigorous process of learning incrementally from uncertain observations of complex phenomena—a process that, as discussed in the next chapter, relies on Bayesian (or related) statistical methods.

The selection of additional biomarkers for use in refining the selection of additional patients to include in the trials will be guided by a mechanistic biochemical understanding of why the biomarkers are relevant, along with the types of data already used by the FDA when licensing orphan drugs. Current laboratory tests, such as those developed to mimic various aspects of the human liver or heart cells, can be used to confirm that a drug can indeed interact with a biomarker in a way likely to affect the drug's performance. The in-depth investigation of the response of individual patients, coupled with today's sophisticated laboratory tests, can do much to ensure that biomarkers are chosen based on objective, scientifically plausible criteria rather than wishful thinking. Moreover, the analytical engines that quantify the strength of drug-biomarker links to clinical effects need not even know whether the effects are medically good or bad—regulators can see to it that the computers wear the blindfolds.

If the analytical engine is doing its job well, the adaptive trial will progressively home in on the taxonomic aspects of the disease—if any—that determine when a drug can perform well down at the molecular and cellular levels, along with biomarkers that determine when it causes unacceptable side effects. The drug's clinical performance should steadily improve as treating doctors gain access to the information they need to predict when the drug will fit the patient. If performance doesn't improve, either the drug or the engine is failing; either way, the trial should end.

And if performance does keep improving? To begin with, the trial can start expanding again—more clinicians can enlist and treat more of the right patients. And if the drug's numbers continue to improve, what next?

One way or another, accelerated approval would be followed by further postlicensing review. This may take the form of conventional blinded trials, but should it? With comprehensive tracking and reporting systems in place, biochemists, unblinded clinicians, and Bayesian engines can continue to develop the patient-selecting biomarkers and rigorously assess their clinical implications for as long as the drug is used. As discussed further in Chapter 11, the FDA already relies on this process to expose rare, long-term side effects that don't surface during the front-end trials.

Adaptive licensing is a necessary corollary to the adaptive and open-ended development of drug-patient science, and formalizing it would also force Washington to be more candid about the scientific realities of the drug licensing process. When, if ever, a drug company should be able to start selling a drug for profit, and for what medical purposes, can be guided and limited by the accuracy of the constantly evolving databases and analytical engines that link known molecular effects to desired clinical effects. But when we feel confident enough to justify using a drug to treat any particular patient or disorder is not a strictly scientific question, and Washington should stop pretending that it is. As the databases grow and the analytical engines improve, the authority to make the final calls should shift progressively from Washington to professional medical associations whose members are engaged in the ongoing battle against a disease, and on down to front-line doctors and patients.

This will leave the FDA with a narrower task—one much more firmly grounded in solid science. So far as efficacy is concerned, the FDA will verify the drug's ability to perform a specific biochemical task in various precisely defined molecular environments. It will evaluate drugs not as cures but as potential tools—molecular scalpels, clamps, sutures, or dressings, to be picked off the shelf and used carefully but flexibly down at the molecular level, where the surgeon's versions of these tools can't reach. The FDA will retain the power to require that the drug be prescribed only by certain specialists, and only to patients who are tested and tracked as may be necessary to ensure that the drug is prescribed in ways consistent with what is known about its effects. If physicians and patients are willing to cooperate, the data gathering and analysis used in adaptive trials can also be used to systematize the essential and rapidly expanding sphere of off-label drug prescription.

Safety is (and will forever remain) a trickier issue than efficacy. All new drugs will continue to be screened at the threshold for acute toxicity before adaptive trials begin. Genetic factors and other biomarkers that are linked to some fairly common side effects, such as those associated with the body's ability to a metabolize a drug, have already been identified, and adaptive trials can search systematically for others. But as the PCAST report notes, other side effects may always be lurking just over the horizon, some of them possibly serious. Some balancing between known benefits and unknown risks will always be required, and the balancing should itself be an ongoing process as clinical experience accumulates. The best that science and regulation can do for the individual patient is provide the best possible estimates about how much confidence can be placed in the personalized prediction made by a well-designed analytical engine. If the drug is effective for some purposes and the analytical engines are doing their job, the drug's overall performance should steadily improve as we refine our ability to link both good effects and bad to patient-specific biomarkers.

Finally, though much of Washington will recoil at the idea, we should conduct systematic comparative effectiveness studies of the regulatory process itself. However the front-end trial is scripted, one of its purposes is to establish a reliable basis for prescribing a drug safely and effectively to future patients. Conventional FDA trials provide a familiar path to that end, relying on human expertise and one specific type of statistical investigation. Adaptive trials that rely on Bayesian analyses of continuously evolving patient databases offer a different path to the same end. Those two alternatives can be tested against each other, as was in fact done in the EuResist "Engine Versus Experts" study—there are systematic ways to find out if adaptive trials used to educate a Bayesian computer can provide better predictive guidance, and provide it sooner, than trials scripted by the FDA, with the details of what was learned collapsed into FDA-approved labels. Clinical experience with a drug that is widely prescribed off label in ways later vindicated in FDA-approved clinical trials offers further opportunities to test how the Bayesian computers measure up against the empirical and analytical methods of the past.

That a drug trial must often begin with an imperfect molecular understanding of a disease's biochemistry also raises a final question of institutional competence. At present, the FDA passes judgment, implicitly or explicitly, on two scientifically distinct issues: a drug's ability to control a molecule down there, and the role that the same molecule plays in causing clinical effects up here. The first obviously involves the drug. But the molecules that precisely define a disease and control its progress are matters of biological science. The FDA has quietly emerged as America's chief taxonomist of health and disease, policing not just drug-disease interactions but also the disease-defining science and all the diagnostic and prognostic measurements used to monitor a disease's advance or retreat inside the individual patient.

But the NIH, not the FDA, is the agency with the deep expertise in diseases, and it is therefore the agency best qualified to decide when specific, measurable, molecular-scale changes in a patient's body have some reasonable prospect of playing a role in changing the trajectory of a disease for the better. The NIH should, at the very least, have independent authority to identify the biomarkers that can play an important role in improving the quality of drug science and the speed at which drugs are licensed. NRC report cochair Dr. Susan Desmond-Hellmann has suggested that biomarker validation might also come from "other regulators or the American Heart Association or the American Cancer Society."

Sooner or later the individual doctor and patient should be added to that list. The accumulation of molecular and clinical data in networks such as the one proposed by the NRC will steadily improve medicine's ability to make an accurate, personal, biomarker-based prognosis of how the untreated disease is likely to progress inside the patient. The doctor and patient will thus gain access to concomitantly accurate estimates for how much benefit the individual patient is likely to derive from drugs that modulate molecules involved in propelling the disease. Together, the patient and doctor will then be better qualified than anyone else to decide when it makes sense to start fighting the clinical future of the disease by using one or more drugs to attack molecular problems here and now.

"IN THE AREA of orphan and rare diseases," the PCAST report notes, the FDA "has demonstrated notable regulatory flexibility and innovation in order to get drugs to very small patient populations who suffer rare conditions." The Orphan Drug Act, as it happens, already implements much of what the PCAST report would extend to many more drugs.

The ODA covers drugs directed at rare diseases, many of which are caused by a single, rare genetic disorder, which points directly to a single protein that an effective drug can target. This makes it easy to frame trials that fit the drug to the right patients from the get-go, and track one key aspect of the drug's performance down at the molecular level. The ODA also gives the FDA the flexibility to license drugs on the strength of individual patient case reports, or even studies conducted in animals or laboratory glassware.

Drugs designed and licensed this way, are, in effect, recognized and used as molecular tool kit drugs from the start. Gleevec, the pioneering orphan drug discussed in Chapter 3, was developed to treat one rare form of leukemia propelled by a single flawed gene found on the rare Philadelphia chromosome. But the drug proved effective against a fistful of other cancers and has been widely prescribed off label. At its peak, little orphan Gleevec was raking in $5 billion a year.

Over half of all certified orphans end up as wards of Pharma, and many end up treating big crowds. The orphan billionaires epitomize the gulf between the old medicine and the new. The government still defines the orphans from the top down. But the biochemists fit drugs to diseases by searching for a common molecular problem lurking underneath, and if the same problem lurks under ten diseases, then Gleevec earns tenfold profits.

The FDA has designated as orphans about seven thousand rare conditions that collectively affect some thirty million Americans, and has approved about 350 orphan drugs. The orphanage currently fosters about one-third of the FDA's successful graduates and is now home to "the most rapidly expanding area of drug development." This is widely viewed as a "roaring success."

Yet only about two hundred of the seven thousand orphan diseases have become treatable in the almost three decades since the ODA was passed. Meanwhile, new links between rare genes and disease are discovered daily. And there are undoubtedly many more orphan diseases out there—older estimates didn't take into account the millions of single-letter genetic variants discovered in the last few years. We will need still more drugs to treat uncommon genetic diseases that aren't quite rare enough to qualify as orphans.

The ODA assumes Washington can spot an orphan disease before biochemists have developed even a single drug to treat it. In fact, molecular medicine often discovers the rare orphans lurking underneath the common clinical symptoms only when it starts developing drugs and cocktails and discovers that each one works in just a small fraction of patients. The FDA should apply the ODA's key licensing provisions to every new drug that promises some biochemically tangible advance in the treatment of an untreatable disorder.

WHAT IT WILL take to get drug companies, doctors, and patients engaged in adaptive trials is a separate question. Experience with HIV and AIDS drugs and an early adaptive trial of a Pfizer drug for acute stroke therapy indicates that patients are considerably more willing to volunteer for trials in which they are guaranteed some kind of treatment than for trials in which they take their chances on the flip of a coin. Drug companies and doctors, however, may hesitate to start prescribing new drugs under less tightly controlled conditions until they are confident that the data acquired will be analyzed using rigorous statistical methods, not cherry-picked in an unscientific search for anecdotes that can be used to condemn a drug at the FDA or launch lawsuits. The FDA side is easily addressed. As we shall see, a law that establishes an alternative compensation system for vaccine-related harms already provides one reasonably fair and accurate alternative to the wildly unpredictable tort system.

The shift to an incremental, adaptive licensing process in which trials and treatment converge and learning continues indefinitely will also require policies that engage market forces in the long-term collection of large amounts of information that isn't currently collected by anyone. Washington alone can't afford to fund this process, and it has strong incentives to restrict expensive tests that might lead to expensive treatment. As discussed in more detail in Chapter 16, market forces can and should be engaged instead, in ways that ensure that the process begins with well-informed consent and is framed to advance the medical science and perhaps improve the patient's own prospects as well.

However framed, adaptive trials and related policies must advance a single, overarching objective: clearing the way for much closer, long-term collaboration between independent doctors and drug designers, most of whom, of course, work for drug companies. Under current rules, the companies can have no active involvement during licensing trials. For the most part, they are likewise barred from dealing directly with doctors or patients to monitor and investigate how their drugs are performing, once licensed. And the companies must constantly worry that if they fail to blame their own drug as soon as they receive any reports of possible problems, they will expose themselves to swarms of trial lawyers alleging cover-up. Medicine cannot develop the integrated drug-patient science on which molecular medicine depends in a regulatory environment that so rigidly separates the biochemists who develop drugs from the doctors who prescribe them and the patients whose bodies contain most of the information the science needs.

REASSEMBLING THE PIECES: PART 2

IN 1948, A century after John Snow tracked cholera to the Broad Street pump and removed the handle, his successors at the NIH began searching for handles that might be removed to quell America's rising epidemic of heart disease. They signed up 5,209 residents of the small town of Framingham, Massachusetts, to participate in a long-term study that would track their cardiovascular health and an array of possible risk factors. But the researchers faced an immediate practical problem: using traditional statistical methods to analyze every possible combination of ten risk factors would have required tracking hundreds of thousands of people to get a sufficient number of representatives of each possible combination.

At about the same time, Jerome Cornfield, one of the NIH's own statisticians, set about rediscovering the genius of Thomas Bayes and Pierre Simon Laplace, the two eighteenth-century fathers of Bayes' theorem. The one-line Bayes formula provides a systematic way to calculate "reverse probability"— how confidently we can attribute an observed effect (lung cancer or a heart attack, for example) to a suspected cause (cigarettes or high cholesterol). Cornfield's landmark 1951 paper demonstrated how statistical methods based on that theorem could be used to establish with high confidence that most lung cancers had been caused by cigarettes. As Sharon McGrayne recounts in her 2011 book, *The Theory That Would Not Die*, Bayes' theorem has emerged in recent years as "arguably the most powerful mechanism ever created for processing data and knowledge." Her book provides a lucid and accessible account of the modern development of Bayesian analysis.

Pinning down reverse probabilities with high confidence is extremely difficult when a single effect might be the product of many causes that occur in different combinations or interact in complex ways. When conventional statistical tools are used, getting robust answers for all possible combinations of all relevant factors requires massive amounts of data. Bayesian statisticians converge on correct answers much more quickly and efficiently by adding science to the analysis in a way that allows them to rely much less heavily on purely statistical correlations.

Richard Wilson, a Harvard professor of physics, provides a simple illustration: how believable is a child's report that says "I saw a dog running down Fifth Avenue"? To answer the question using conventional "frequentist" tools, one might conduct a study of children randomly assigned to walk a path with Fifth Avenue–like pedestrian traffic and distractions for all, but a dog briefly included only half the time. Statisticians can tell us how many children would have to be tested to arrive at a statistically reliable measure of how much we can trust such reports, assuming that all the factors that they can't control for—the child's eyesight, veracity, yearning for a puppy of his or her very own, and so on—are randomly distributed among children. Dog size may be a factor, too, so if FDA statisticians were in charge, they would want a representative mix of breeds, from Great Danes to Chihuahuas. Reports of a lion sighted on Fifth Avenue would require new trials with the right mix of lions. Same with stegosaurus reports. The FDA could handle them all, so long as someone was willing to pay for each trial.

A Bayesian, however, would start at a different point and arrive at reliable answers much faster. We are dealing here with a typical reverse probability problem: we have an observed effect—the child's chatter—and we're wondering how confidently we can attribute it to the suspect cause. But we're talking Fifth Avenue, where dogs are quite common. Accepting an "I saw a lion" report requires additional information: were the Ringling Brothers in town, and did their truck crash? "I saw a stegosaurus" is never believable, not even if Steven Spielberg is in town. The reliability of each report depends not only on the child but also on knowledge that has nothing to do with the child at all—knowledge about where lions roam and dinosaurs don't.

Bayes' theorem provides a systematic, rigorous way to insert that kind of external knowledge into the analysis when calculating reverse probability. Indeed, the rise of modern Bayesian analysis began with the recognition that "statistics should be more closely entwined with science than with mathematics." As one Bayesian analyst put it: "The limit of [frequentist] approaches just isn't obvious until you actually have to make some decisions. You have to be able to ask, 'What are the alternative states of nature, and how much do I believe they're true?' [Frequentists] can't ask that question. Bayesians, on the other hand, can compare hypotheses."

We already have good numbers for many of the alternative states of nature on Fifth Avenue, and if we didn't, we could acquire them without conducting a long series of double-blind trials. A child's report that the taxi he saw was yellow is quite believable; pink, not so much. The example often used in medical textbooks addresses the use of mammograms in the routine screening of forty-year-old women: the results have an 80-20-10 accuracy rate, spotting 80 percent of the tumors and missing 20 percent, with the 10 percent being "false positive" reports of a tumor that isn't there. So these mammogram reports are of course wrong 97 percent of the time—twenty-nine out of every thirty frightened patients that they send scurrying for a biopsy or some other test don't need it. If you have no idea where that "of course" came from, and don't believe it, you're in good company—surveys indicate that many American doctors don't, either. But for Bayesians, this is a very simple calculation.* Mammograms are usually wrong not because radiologists are incompetent but because breast cancer is rare—more lion than dog. When used in routine screening for rare diseases, any test that is even a bit less than perfect will report many more false positives than true positives, because the number of healthy individuals screened will dwarf the number who are sick.

Cornfield had helped design the Framingham heart disease study in 1948; a decade later it still hadn't lasted long enough, nor was it large enough, to pin down any risk factor with high confidence. But using Bayesian analysis, statisticians can refine probabilities as fast as new evidence is acquired, and in the search for rare causes we often acquire information about suspects that *don't* matter much more quickly than information about those that do. In the first decade of the study, 92 of 1,329 adult males had experienced a heart attack or serious chest pain. Based on a Bayesian analysis of the various combinations of risk factors presented by those who had and hadn't, Cornfield was able to reframe the study around just four risk factors—cholesterol, smoking, heart abnormalities, and blood pressure. The "multiple logistic risk function" he developed has been called "one of epidemiology's greatest methodologies."

Using the limited amounts of data obtained during the early phases of a trial to narrow the trial's focus is *not* the same as concluding that we are now highly confident that we know which risk factors matter; we are just more confident than we were a while ago, we can calculate how much more, and when the numbers look encouraging enough, we can focus more of our attention on some factors and less on others. Later results may either reinforce that early confidence boost or undermine it—the process is self-correcting. And it worked well in Framingham. Using data acquired in the early years of the study to narrow the range of what remained to be

---

*Among forty-year-old women picked at random, about 40 out of 10,000 have breast cancer, and 9,960 don't. Radiologists with an 80-20-10 skill at reading mammograms will detect 32 of the 40 actual cancers, and miss 8. They will also suspect cancer in 996 women who don't have it. So, of the 1,028 positive calls made, 32 (about 3 percent) will be correct.

explored statistically, Cornfield hastened the arrival of statistically robust correlations that have since helped save millions of lives.

———————

BIOLOGISTS NOW RELY almost entirely on Bayes or closely related analytical tools to track complex diseases, or their absence, back to genetic and other molecular factors. Chapters 4 and 5 include illustrations of the probabilistic causal networks that emerge when modern Bayesians use powerful computers to analyze large amounts of detailed molecular and clinical data. As those analyses demonstrate, Bayesians can begin with large numbers of suspect biomarkers, each one linked to all the others—the strength of each link based on such things as biochemical logic, laboratory experiments, and experience with other diseases, and often not much better than a guess— and then systematically adjust all the numbers until they align with all the available data on combinations of biomarkers that were present or absent in patients who did or didn't develop the disease. With enough data to analyze, the biomarkers that play no role will drop out of the picture. Those most strongly associated with the disease will, in one typical graphical representation, migrate toward the center, closest to the point that represents the disease itself.

The same Bayesian statistical methods can likewise be used to analyze links between unusually good health and the underlying causes—genes, for example—that keep some heavy smokers cancer free, or allow some patients to control HIV on their own. And they can add lung cancer drugs and various measures of a cancer's advance or retreat to their analyses as well.

With drugs, a series of frequentist trials can eventually yield the same answers as adaptive Bayesian trials—each separate trial will test a different combination of suspect causes in a suitably large number of patients, and when every combination of biomarkers has been tested, we will be statistically confident that we know how likely it is that the drug's good or bad effects can be attributed to each combination. But if the disease is biochemically complex, a great deal of time and expense will be dedicated to testing suspect causes that don't play any role at all. When Cornfield first worked with researchers who were testing drugs in laboratory animals, the rigidity of the protocols appalled him. The trials plodded forward according to script long after a Bayesian would have concluded that the starting hypothesis had been disproved.

The FDA itself is sometimes a closet Bayesian, because at times there is no other ethical or practical option. A separate team of unblinded doctors typically monitors the results of clinical trials from a distance and can halt the trial if the results seem so clearly good or bad that continuing the trial would be unethical; the trial of AZT, the first HIV drug, ended that way. And the identification of drug side effects that aren't bad enough to halt a trial invariably involves an ad hoc Bayesian search for patterns that suggest the drug is to blame, though they often don't pin down the link with high statistical confidence. If the drug gets licensed, the label will warn doctors to be on the lookout for such effects, and the agency has in place various processes for collecting reports of other side effects that the drug may have caused thereafter. But ad hoc Bayes is a far cry from the real thing, and most of the statistically robust molecular-based safety science emerges, if ever, after the drug is licensed. In the words of one prominent expert in pharmacoepidemiology, "Evaluation of drug safety has much in common with evaluation of a patient, in that both are inherently Bayesian processes. Armed with an informed set of prior probabilities, one looks for signals. Suggestive pieces of evidence are then worked up further, even if they do not initially offer black-and-white confirmation of 'significance.' Additional targeted studies are then conducted in a timely way to follow up on promising hypotheses."

During the licensing trials themselves, the FDA's frequentist protocols bar ongoing monitoring to search for reasons different groups of patients are responding differently and then adjust treatments on the fly.* The FDA's "controlled" trials thus deliberately exclude controls that unblinded doctors might otherwise develop and use to guide the inclusion of new patients and the exclusion of older ones as the trial progresses.

Instead, the FDA's frequentist statistical methods consign to chance everything that isn't understood and controlled at the outset and let statistical analysis take it from there. These methods assume that when a drug lands inside a human body anything is possible, but some things are just less likely than others; assuming a specific probability distribution for the limitless number of unknown drug-patient interactions keeps the statistics and the trials manageable. But while new drugs can surprise us in many unanticipated ways, biochemistry is not a realm in which anything is possible. How drugs and human bodies interact is constrained by solid rules of biochemical science, and we now have the power to identify those constraints, molecule by molecule, thus reducing our need to rely on blind statistics.

The alternative states of nature that can affect a drug's performance are defined mainly by the biomarkers that can interact with the drug in all the different combinations that occur in patients who use it. A clinical trial of a drug that targets a biochemically complex disease will always begin with

---

* FDA protocols do allow "subgroup analysis" of the results at the conclusion of some trials, but only using statistical analyses that are heavily stacked against approving the drug.

an uncertain and incomplete understanding of the drug-biomarker science. Bayesian choreographers of clinical trials can deal with many suspect bio-markers and recalculate the strength of the links among drug-biomarker clusters and various measures of the patient's health as fast as they acquire data about how different patients respond well or badly. Bayesians can like-wise deal with complex, multidrug regimens from the start, and continue refining them forever.

They can, for example, incorporate what science has long known—or just found out—about how different breast cancer or HIV molecular recep-tors affect a drug's performance, or about how fast cancers or HIV infec-tions mutate at different stages of their assault on our bodies. The EuResist analytical engine also takes into account the fact that three main classes of HIV drugs are used to target three different aspects of HIV's chemistry. Bayesians can start quantifying the likelihood that a new drug will perform well as soon as *any* possibly relevant biochemical information is acquired. They can begin with evidence acquired in glassware and test animals. As we have seen (Chapter 5), they can start quantifying the "biochemical beauty" of a drug—the likelihood that it will successfully reach its intended target without causing side effects—by considering the experience gained with other chemically similar drugs.

None of these sources of data can finish the job. But they can help launch an efficient, robust, self-correcting process that can, as it tracks a drug's ef-fects across biochemical space and time, steadily improve our confidence in our ability to select the patients in whom the drug will perform well. Bayes-ians need not select some arbitrary number of patients to be tested in a trial that will end at some arbitrary point in time. However simple or complex the disorder, the accumulation of valuable data can—and should—continue for as long as the drug is prescribed, because every patient is unique and pa-tient chemistry doesn't stand still, not even in a single body.

In the early stages of a drug trial, the negative information will be more valuable than the positive. The negative data points are the ones that al-low the trial to home in on the molecules that do matter, and then use that information to select for the next round of the trial patients who are more likely to have a positive outcome. As data accumulate, multipatient analyses expose the patterns that can be used to understand the implica-tions of the massive amount of data extracted from a single patient, spot molecular changes that foreshadow clinical problems, and guide customized treatment.

This is the process that will systematically develop the science that ul-timately matters—the complex, data-rich, integrated drug-patient science. As the FDA's own Dr. Janet Woodcock put it in 2004, drug science is at best a "'progressive reduction of uncertainty' about effects—or 'increasing level of confidence' about outcomes" that comes with the development of "multidimensional" databases and "composite" measurements of outcomes.

———————

THE COMPREHENSIVE TWO-YEAR iPOP monitoring of Michael Snyder (Chapter 2) tracked some forty thousand biomarkers and acquired gigabytes of data. The information gleaned from one patient has little value on its own, other than to that one patient. It is by feeding the iPOP data from many individual patients into EuResist-scale analytical engines that we move progressively toward—though never quite reach—a truly rigorous science of personalized medicine. As of 2011, the EuResist engine was powered by records that span tens of thousands of patients, more than a hundred thousand treatment regimens, and more than a million records of viral genetic sequences, viral loads, and white blood cell counts.

The calculations required to extract cause-and-effect patterns from such huge volumes of complex data are extraordinarily difficult—so difficult that an appreciation of the true power of Bayesian analysis had to await the digital revolution. Today's computers, however, routinely perform half a million interdependent causal calculations in parallel—a power almost inconceivable even a decade ago. The digital wizards, as it happened, needed the power themselves—their devices and networks are constantly racing to link what matters to you right now with just the right puff of data stored somewhere in the vast digital cloud that surrounds you. Doing that efficiently is essential, which means anticipating what you want before you ask for it—something digital Bayesians figure out how to do by learning from experience about the alternative states of nature commonly found in your microprocessor or brain. Bill Gates has attributed much of Microsoft's success to its mastery of Bayesian networks. Bayes is undoubtedly powering Google, Facebook, and almost all other networks that tout their ability to connect you with the right him or her—or, failing that, the right super-cheesy pizza.

The digital community has also grasped—far out ahead of the FDA and much of the medical community—how fast molecular medicine can now advance by taking full advantage of the recent convergence of astonishingly powerful biochemical diagnostic tools and digital technology. Andy Grove, the pioneering founder and for many years CEO of Intel, took on the FDA in a *Science* magazine op-ed in late 2011. The biotech industry, he notes, currently spends more than $50 billion a year on research that yields some twenty new drugs. Almost half of all clinical trials are delayed by difficulties in convening a group of patients that meets Washington's rigid requirements, and trial costs spiral upward as a result. The FDA's trial protocols were conceived at a time when science lacked the computing power that it can draw on today. That power now makes possible a process that, in its early phases, enlists patients much more flexibly, to "provide insights into the factors that determine . . . how individuals or subgroups respond to the drug, . . . facilitate such comparisons at incredible speeds, . . . quickly highlight negative results, . . . [and] liberate drugs from the tyranny of the averages that characterize [FDA-scripted] trial information today. . . . As the patient population in the database grows and time passes, analysis of the data would also provide the information needed to conduct post-marketing studies and comparative effectiveness research."

A year later, 23andMe, the Google-connected company that provides consumer genetic sequencing services, announced it would allow other providers and software services to develop applications that would interact with the data entrusted to 23andMe by its customers. Hundreds soon did. Their interests, *Wired* reported, include "integrating genetic data with electronic health records for studies at major research centers and . . . building consumer-health applications focused on diet, nutrition and sleep." For individuals, 23andMe's platform will, in the words of the company's director of engineering, serve as "an operating system for your genome, a way that you can authorize what happens with your genome online."

Meanwhile, Washington remains focused on why ordinary citizens should *not* be permitted to read their own biochemical scripts. As we have seen, the FDA is determined to protect us from reports provided by diagnostic sniffers or companies like 23andMe that, however biochemically accurate they may be, might lead to "unsupported clinical interpretations"—just as routine mammograms often do.

Dollar doctors believe in Bayesian calculations, too, and invoke them to save money. In 2009, Bayes-based reasoning helped persuade Washington's Preventive Services Task Force to recommend against routine mammograms for women in their forties. Such recommendations make sense for the statistically average woman, but not for individual patients and their doctors. A basic patient interview that looks at family history (which of course points to genetic factors), obesity, smoking, drinking, and having a first pregnancy later in life can identify factors that may tip the scales back in favor of early screening. And molecule-spotting diagnostic sniffers can quickly transform probabilities into near certainties—here again, Bayesian calculations can tell us how near.

Which gets us back to the government's aversion to letting the masses sniff away at their own biomarkers until Washington is confident that it understands their clinical implications. Getting there depends in large part on doing exactly what 23andMe wants to help lots of people start doing today: feed torrents of biochemical data into the rapidly expanding cloud of digital-biochemical-clinical data, to be tracked by Bayesian engines that will progressively refine our understanding of all the factors that don't matter, to isolate those that do.

When 23andMe and others let the rest of us catch up with the iPOPing professors at Stanford and gain easy access to the digital power that can discern the causal patterns in the torrents, the first thing we should do is establish a baseline profile of our excellent health, and keep it up to date thereafter. With that information securely stored and pooled with enough data from other patients, the Bayesian engines will take it from there. When we suddenly find ourselves diabetic, they will probably be able to tell us whether a viral infection, a bad diet, or some other factor was to blame. When we try a potential cure, we will be able to track and at least tentatively evaluate its efficacy almost immediately, down at the molecular level. To establish a control baseline for its drug crowd science, the FDA directs doctors to prescribe placebos, but as Stanford's Professor Snyder noted, the patient's own healthy, unmedicated history can provide the best possible control for tracking a disease to its root cause and evaluating the drug prescribed to cure it.

Why isn't the FDA already on board? It accepts Bayesian methods when licensing devices—lenses, implants, artificial hips and such, and sniffers, too. In February 2010, it did finally issue a draft guidance for adaptive drug trials, and as noted earlier, the FDA has taken a few small, hesitant steps that point to the possibility of a fundamental shift in the way it will script clinical trials and pass judgment on drug science. But it has clearly failed to proceed at the pace that many outside experts have been advocating for years. The PCAST report notes "widespread concern that scientific discoveries are not being translated rapidly enough into urgently needed medicines for patients."

One practical problem, the PCAST report points out, is that the FDA's "incompatible" and "outdated" IT systems are "woefully inadequate." The agency lacks the "ability to integrate, manage, and analyze data . . . across offices and divisions," and the processing of a new drug submission may thus involve "significant manual data manipulation." To implement the PCAST recommendations, "the FDA will need to fundamentally rethink its information architecture."

One of the FDA's legitimate technical concerns is, apparently, the "prior probabilities" Bayesian analysis requires. In deciding what to make of reports from children or radiologists, or from doctors engaged in a drug trial, Bayesian analysts need estimates of how often lions or women with breast cancer stroll down Fifth Avenue, or how strongly a suspect biomarker affects the drug's performance. These estimates can affect how quickly a Bayesian analysis will converge on a reliable answer, and drug trials must often begin with speculative estimates—too speculative, the FDA worries—of how various biomarkers might affect a drug's performance. The FDA, however, begins with initial guesses, too—about how many patients must be tested for how long to expose enough detail about our complex biochemical diversity. The main difference is that the FDA buries its estimates in trial protocols and reductionist unscientific pronouncements about "safe" and "effective" for the crowd. There is, of course, only one reality out there, and if the drug is prescribed to enough patients, the Bayesian and frequentist analyses of the results will invariably converge on the same understanding of how a drug's clinical effects are shaped by the various biomarker combinations presented by different patients. Without enough data, both can make mistakes. The Bayesians will correct theirs much faster and at lower cost, while delivering better treatment to more patients along the way.

Bayesian analyses look messy mainly because they dare to deal forthrightly with complexity. Bayesians don't wear blindfolds. But they don't choke when biochemical reality gets complex, either. "Far better an approximate answer to the right question," as one Bayesian put it, "than an exact answer to the wrong question."

12

ANTHRAXING WALL STREET

THEY'RE NORMALLY STICKY and easily contained, but when coated with silica, the "weaponized" spores disperse through the air like ragweed pollen in spring. An added advantage is that the disease they cause isn't contagious. They can be used to kill New York, say, without unleashing an epidemic that might spread to Islamabad. They start germinating almost as soon as they enter your body. Bacterial toxins drill their way through your flesh, gangrene sets in, organs collapse, lungs fill with fluid, and you die of pneumonia. Anthrax thus delivers in ten days what HIV delivers in ten years.

We will find out just how far we have slouched down the road to pharmaceutical serfdom on one windless summer evening when a low-flying Cessna scatters two hundred pounds of anthrax over the rooftops of Manhattan from Harlem to Battery Park. One Cessna's worth could easily kill a million people—Wall Street and Broadway, Mayflower WASPs and rainbow immigrants, Lower East Side and Upper West, people of every hue, ethnicity, gender, lifestyle, and sexual orientation. The threat, numerous intelligence assessments have concluded, is all too real.

If you spot the plane over Times Square tomorrow, prepare for a shot of BioThrax—a vaccine developed for the Pentagon in the 1950s—and prepare, too, for some side effects, unpleasant or possibly worse. Or scramble to find an antibiotic that might work. Or head for the hills, because the plane may instead be carrying Ebola or one of a fistful of other killers for which we have no antidotes at all, or perhaps smallpox or plague genetically rejiggered to evade the antidotes that we've got. In today's ongoing struggle between designed germs and designed drugs, we aren't ready to defend ourselves against existing threats—nowhere close. And we aren't going to close the gap anytime soon.

That should surprise many and will leave the rest of us feeling savagely wise. The germ-fighting side of the drug business—all of it, not just the counterterrorism part—is already pretty much where the health care left says the whole pharmaceutical industry belongs. Vaccine manufacturers supply what Washington requisitions, at prices dictated and quality minutely controlled by one omnipotent customer—if the right companies show up to supply it. But they no longer do, for reasons that will soon become clear. The manufacturers of antibiotics fare only slightly better, and most companies have solved the problem in the same way, by abandoning the market.

Yet it is in these two sectors of the pharmaceutical industry that we most need constant innovation spurred by vigorous competition among drug companies, because here they are competing mainly against nimble, adaptable, and relentlessly competitive killers. As we have seen, the dynamic biochemical complexity of our own bodies often presents similar challenges. We should be reversing the policies that brought us to this point, not extending them to the rest of the industry, as some aspire to do in implementing the Affordable Care Act.

DRUG-RESISTANT DISEASES now kill an estimated sixty-five thousand Americans a year, more than breast and prostate cancer combined. In 2005, according to one estimate, about nineteen thousand of them—slightly more than were killed by HIV—succumbed to a methicillin-resistant staph bacterium. That we are now better at subduing nimble HIV than we are at stopping a big fat bacterium reveals much about the economics of germ-killing drugs.

With HIV, Washington inadvertently got the economic fundamentals—as viewed from Wall Street—exactly right. It failed to notice the stealthy virus during the first decade or so of its spread across America, so once drug companies grasped how widely it had spread, they had plenty of incentive to search for a cure. Quite a few patients were already in the late stages of AIDS, or soon would be, which made it relatively easy for drug companies to persuade the FDA that AZT and subsequent drugs could help. The U.S. vaccine market was also on the verge of collapse—brought to that point, as discussed shortly, with much help from Washington—so help clearly wasn't arriving from that quarter. And no sexually transmitted disease has ever been eradicated by culture warriors or condoms. About 1.1 million Americans are now HIV positive, and about 50,000 are still infected every year. In sum, the sudden discovery of a widespread, lethal epidemic, and Washington's persistent failure to beat it, kept one small corner of Washington's germ-fighting bureaucracy safe and effective for Wall Street.

But not the rest. While HIV was spreading unnoticed across the country, Washington and the trial lawyers were also entrenching themselves as the final arbiters of vaccine and antibiotic science. In 1972, Congress had transferred responsibility for vaccines from the Biologics Lab to the FDA. The FDA had spent the previous decade writing new rules in the shadow of thalidomide, and safety was its paramount concern. With vaccines, which are administered to huge numbers of healthy people, that meant conducting bigger and longer front-end trials. A recently licensed vaccine—it protects against a common rotavirus that kills about sixty American children a year and hospitalizes fifty thousand—required ten years of development, followed by another sixteen years of tests involving seventy-two thousand patients.

When HIV surfaced, the FDA grew very worried about the possibility that other slow-moving, undetected viral killers might be lurking in the live animals, eggs, and cell cultures that are used to mass-produce vaccines and monoclonal antibody drugs. It was a reasonable concern, because viruses are difficult to detect. But the regulation of vaccine manufacturing grew increasingly stringent and rigid. It has ended up locking much of the manufacturing into comparatively clumsy and expensive germ-farming methods of the 1950s. Reliance on those now outdated technologies significantly slows the process and raises the cost of moving from the design of a new vaccine to mass production. Much more advanced methods are used to mass-produce the mAbs that help treat cancers and other diseases, and costs there have fallen steadily.

Meanwhile, the toxic legal spores released in the polio lawsuits of the 1950s and 1960s had mushroomed. With help from such authorities as the *National Enquirer* and Oprah Winfrey, junk science, often promoted by trial lawyers, periodically turned into runaway litigation in which vaccines were held responsible for harms they hadn't caused. A deeply flawed study published in 1977 in the *Lancet*, a prestigious British medical journal, led to a cascade of suits that blamed the whooping cough vaccine for brain damage, unexplained coma, Reye's syndrome, epilepsy, sudden infant death, and other afflictions. It took years of follow-up studies to exculpate the vaccine. Another study, published in 1998 in the same journal—this one involving the measles vaccine and later characterized as having involved downright fraudulent data—unleashed a tsunami of U.S. litigation blaming autism on a mercury-based compound that had long been added to many vaccines to provide protection against the risk of contamination by fungi and bacteria. Twelve years after it helped launch this debacle, the *Lancet* conceded that it had all been a big mistake.

Midway through the industry's collapse, Washington recognized that it had a problem. In 1985—just as the tools of intelligent design were coming of age and drug companies were poised to launch their astonishingly fast and successful assault on HIV—the health arm of the National Academy of Sciences concluded that innovation in the vaccine industry was grinding to a halt. Nobody paid any attention. Ever since, the raw cost of identifying targets and designing and mass-producing biochemically potent molecules has been plummeting—but the total cost of developing new vaccines has risen relentlessly.

The explosion of vaccine litigation was the one aspect of the industry's economics that Washington couldn't ignore. When an alarming new swine flu surfaced in 1976, Washington placed a big order for a vaccine; insurers, worried about lawsuits, refused to cover the drug companies, which then refused to supply the vaccine. So Washington wrote the coverage itself and wound up fighting claims for the next decade, paying out far more than it had anticipated. A decade later, lawsuits had sent vaccine prices soaring across the board, and as the largest buyer, Washington was indirectly footing most of the legal bills. Plaintiffs were also suing Washington itself for failing to enforce its own rules carefully enough to prevent the harms that vaccines supposedly caused. The Supreme Court upheld their right to do so.

In 1986, Washington imposed an excise tax on vaccines and established a fund to provide compensation for injuries caused by children's vaccines—with expert panels, not juries, deciding all questions of cause. Trial lawyers continued filing cases and persuaded some state judges that the federal law said they still could. In early 2011, in a case involving a vaccine administered nineteen years earlier, the Supreme Court finally made it fairly clear that they couldn't.

Meanwhile, Washington had established itself as the dominant purchaser and distributor of vaccines as well. First drafted following a measles epidemic in 1989–91 and finally enacted in late 1994, the Vaccines for Children program was intended to serve as the first step in the Clinton administration's plan to establish a comprehensive national health care system. Today Washington directly or indirectly pays for well over half of all the vaccines we use, and it effectively controls prices, profits, and which vaccines are produced by which companies.

As Washington took control of almost all the science and most of the money, manufacturers were left to curry political favor and chase the economies of scale involved in manufacturing huge quantities of a few dozen tried-and-true products. For Wall Street, vaccines thus came to present an economic vise of highly uncertain but potentially staggering costs. There was the cost of dealing with Washington for a decade or more up front, with no certainty of securing a license at the end, and the cost of dealing with tort lawyers indefinitely into the future. Manufacturing costs depended in large part on Washington's directives. The price of the vaccine, if approved, would be determined largely by negotiations with a single huge buyer—Washington again. And in lawsuits filed years after the vaccine was administered, with sympathetic plaintiffs squaring off against a huge drug company, anything could happen.

One by one, smaller manufacturers concluded that this wasn't an environment in which they were likely to prosper. More than a dozen companies were manufacturing childhood vaccines in the 1950s; by 1986, almost all of them had abandoned the market, and as they did they surrendered their licenses. Some 380 vaccines had FDA licenses in 1967; by 1984, the number had dropped to 88, and by 2004, there would be only a few dozen. In 1957, five important vaccines were being supplied by twenty-six companies; by 2006, only four major companies would be making any vaccines at all in the United States. According to one recent count, more than half of the routine children's vaccines now come from just one manufacturer, and four others from just two. Four companies dominate the global market.

In 2004, Dr. Scott Gottlieb, soon to be appointed deputy commissioner of the FDA, observed that "most of the vaccine industry long ago ceased to behave as a competitive industry focused on creating better products and being able to earn profits through that innovation." In many cases, he noted, "vaccine makers are constrained to producing largely the same vaccines according to government specs, using manufacturing processes that have been held in place for years by regulations, and selling their finished product mostly for a single government price." And we have witnessed the "near-dissolution of a once pioneering industry."

THE MARKET FOR antibiotics collapsed during the same period, and for many of the same reasons. A 2002 letter to a leading medical journal on infectious diseases, coauthored by a doctor at the Harvard Medical School, blamed the FDA's paralytic oversight over two decades for "the end of antibiotics." Increasingly stringent demands for proof of efficacy, framed in ways that seem "innocent and technical," the authors said, have thrown the industry "into a panic." They have "wreaked irreparable damage to our ability to provide a reliable pipeline of new antibiotics for treatment of serious infections." Things haven't improved since. "Without significant changes from the FDA and perhaps from Congress, the lack of new antibiotics can only be expected to worsen," David Shlaes writes in his 2010 book *Antibiotics: The Perfect Storm*. The FDA's protocols "make it difficult, and at times impossible, to actually carry out the proposed trials."

The FDA did indeed play a role. But most of the credit for this disaster belongs to the single-minded promoters of universal care, and their role must be considered first. In Washington and around the world, their well-intentioned but misguided efforts have vigorously promoted the right to transform potent antibiotics into sugar pills.

At the end of World War II, the federal drug law was amended to bring insulin, penicillin, and then all antibiotics under FDA control. Under a rule that Congress wouldn't extend to other drugs until the 1980s, a copy-cat antibiotic would receive a license simply by showing that it was "bio-equivalent" to the licensed pioneer, with no clinical trials required. This makes copying much cheaper than pioneering and allows cheap copies to flood the market as soon as patents expire.

In the developing world, the cheap copies arrive even sooner. International health organizations obtain huge quantities of powerful antibiotics from drug companies at cut-rate prices, corrupt public health officials often take charge of their distribution, and the drugs end up being sold indiscriminately on the street. Pirates soon start churning out cheap clones. While Pfizer was still charging Americans $10 to $30 per capsule of fluconazole, which treats cryptococcal meningitis, knock-offs were being sold in India and Thailand for about 30¢ apiece. Few protest; after all, who would stand in the way of universal access to life-saving medicines?

But the forces pushing for the indiscriminate distribution of cheap antibiotics ignore an enormous problem: the only way to prevent, or at least significantly delay, the development of antibiotic-resistant germs is to maintain tight control over how antibiotics are prescribed and used. Some governments and societies manage antibiotics well, but most don't. In Norway, only a tiny number of the many strains of bacteria that cause septicemia are resistant to more than one drug; in Greece, over half are resistant. So while Washington has spent much of the last sixty years making the indiscriminate use of antibiotics easier, it has simultaneously made it easier for germs to learn how to dodge them. In the United States penicillin can no longer destroy three-quarters of its former targets. A steadily growing number of diseases once thought to be all but eradicated are now caused by microbial strains that are perilously close to becoming resistant to all our antibiotics. As Steve Jones notes in *Darwin's Ghost*, the germs first learn how to beat our drugs in places where infectious diseases are common and "antibiotics have become a human right." Global travel and trade then ensure the microbes' rapid spread.

Now back to the FDA. The agency worries that its own licenses, if granted too easily, might become part of the problem. If too few infected patients are tested, statistical anomalies might allow an inferior antibiotic to get licensed, which might then become the benchmark for a third, even worse antibiotic, and so on until the industry slouches its way down to licensed sugar pills while the target germs grow steadily more robust. The FDA calls this "biocreep," and in the abstract it is a legitimate concern. To earn its license, a new antibiotic must therefore demonstrate convincingly that it performs as well as or better than those already on the shelf. In the balance between abstract concerns about a new antibiotic that isn't much needed today against the peril of a highly drug-resistant microbe that may or may not emerge some years hence, today's concerns always prevail.

But the comparative safety and efficacy of old and new antibiotics change over time, hinging on how they are used and misused, on label and off, in the United States and abroad. So in a global market that pits antibiotics against germs, we wind up with Washington trying to control drug-versus-drug competition under reactionary rules that keep the germs well out ahead. It's unethical to test a new drug in seriously ill patients when an old, licensed, pretty good alternative is available. Medicine discovers that a drug-resistant strain of an old germ has emerged only when the old treatments fail to beat it. And as the FDA currently operates, we need germs thriving in lots of sick patients to provide the evidence needed to license the drug that will wipe them out.

As occurred in the market for vaccines, most of the manufacturers once actively involved in the antibiotic market have concluded that the profits lie elsewhere. Following the earlier development of penicillin and sulfa drugs, fourteen different biochemical classes of antibiotics were developed between 1940 and the late 1960s. Then, for a long stretch that began soon after the passage of the thalidomide drug law and began to end only when the epidemic of secondary infections associated with HIV emerged in the 1990s, the industry stood still.

Encouraged by the resurgent germs, the antibiotic industry has shown some recent signs of revival. But by all indications, the antibiotics continue to fall further behind the germs. In early 2012 the CDC warned that a strain of gonorrhea resistant to the last line of effective antibiotics is now spreading in the western United States. Infections can still be controlled with cocktails of antibiotics, but the bacterium will continue to mutate and the CDC now sees "treatment failures on the horizon." No suitable antidote is currently in the pipeline. "This is just one more example of a bigger problem—bacteria are developing resistance faster than we're inventing new medicines to fight them," says an antibiotic researcher at the

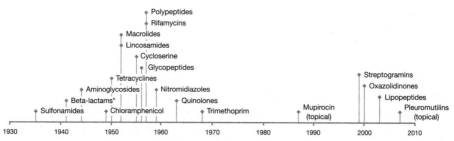

* Beta-lactams include three groups sometimes identified as separate classes: penicillins, cephalosporins, and carbapenems.

FIGURE 12.1   **Fourteen classes of antibiotics were introduced for human use between 1935 and 1968; since then, five have been introduced.**
Source: Resources for the Future, "Policy Brief 6: The Antibiotic Pipeline," May 2008.

Pew Charitable Trusts. A gonorrhea vaccine "remains key to prevention and control," says the CDC, but that is a "distant goal."

WE CAN THANK human terrorists for having recently exposed how poorly we respond when the markets for the antidotes to microbial terrorists collapse.

BioThrax, the anthrax vaccine developed for the Pentagon sixty years ago, was originally tested in a few textile workers, licensed by the Biologics Lab, and then manufactured by a facility nominally operated by the state of Michigan. When the FDA took charge of vaccine licensing in 1972, it initiated a review of all licenses previously issued by the Biologics Lab. But at the FDA no drug could be deemed safe or effective without extensive human trials. That presented a problem: tomorrow we may have a million people exposed to anthrax, and ad hoc testing of any available vaccine will then get started on the double, but today we have none.

So Washington dithered—for the next thirty years. An FDA panel reported that BioThrax was fine, but only for use in the "limited circumstances" of exposure in research labs and from the industrial handling of animal hair—goat hair especially. The agency invited public comment, then simply neglected to finalize the relicensing paperwork. The Pentagon went ahead anyway and BioThraxed about 150,000 men and women involved in the first Gulf War. Some veterans then blamed unexplained illnesses on the vaccine. The FDA suspended production at the Michigan facility. The facility was transferred to private investors with Pentagon connections. The White House said soldiers didn't have to get the shot until the FDA paperwork had been finalized. Lawyers representing unnamed military personnel asked the FDA to admit that it hadn't been.

Then in September and October 2001 letters containing anthrax spores were mailed to national news networks, several newspapers, and two U.S. senators. Of the eighteen Americans known to have been infected by inhaling anthrax spores between 1900 and September 2001, sixteen had died. The letter attacks infected eleven more. The six who surprised almost everyone by not dying owed their lives to several powerful antibiotics—most notably Cipro.

Cipro had been the best-selling antibiotic in the world in 1999, generating about $1 billion in sales for Bayer. The drug was so profitable because it was still young enough to be under patent. At the Pentagon's request, Bayer then got Cipro tested and licensed as the first-ever bioterror drug—for use against airborne anthrax. The Pentagon chose Cipro in the hope that it was still new enough to be out ahead of terrorist schemes to develop antibiotic-resistant anthrax.

After the 2001 attacks, however, three manufacturers immediately offered to supply generic Cipro at a fraction of Bayer's price, and Washington immediately began threatening to let them. Bayer ended up giving away two million tablets and providing another one hundred million at a steep discount. The Canadian government initiated its price negotiations by announcing it would ignore Bayer's patents and order a million tablets of a Cipro generic from another company.

But antibodies work better than antibiotics, so with the challenge to the vaccine's license still pending at the FDA, BioThrax was offered and administered to some Capitol Hill and postal employees. Congress leaped into action and enacted a sweeping new drug law—call it the "Cessna law"—for bioterror antidotes. The law includes a new turbocharged form of accelerated approval. Once a cabinet secretary declares a health emergency—no need to wait for the anthrax-bearing Cessna to arrive first, and the emergency may last forever—the FDA may ignore all its usual rules. No human trials are required, and a "lower level of evidence" applies. "May be effective" is good enough, so long as the "known and potential" benefits appear to outweigh the risks. The decision may be made "within a matter of hours or days." If the drug is also certified as an antiterror technology, any subsequent lawsuits must be filed against the government, not the manufacturer. A federal judge, not a jury, will decide the case, applying evenhanded rules that make plaintiffs' lawyers weep. Awards are capped. No punitive damages may be awarded.

As the ink was drying on the new law, a federal judge agreed that BioThrax still lacked a valid ticket for fending off inhaled anthrax. The FDA immediately issued one. Nothing doing, responded the judge—the agency had evaluated BioThrax versus goats but not Cessnas. So the Pentagon invoked the Cessna law, and in a classic passing of the buck, the FDA licensed BioThrax for use by any and all recipients, military or civilian, so long as they were designated, warned, vaccinated, monitored, and tracked by the Pentagon. A while later, the FDA took care of the goat-not-Cessna problem, completed the paperwork, and gave BioThrax a regular license. But then the agency quietly issued another emergency-use license to bypass some inconvenient fine print in the license of an antianthrax antibiotic. At the same time, ever so quietly, blanket liability protection was also extended to all "activities related to developing, manufacturing, distributing, prescribing, dispensing, administering and using anthrax countermeasures in preparation for, and in response to, a potential anthrax attack."

Congress neglected, however, to pass a law resurrecting our once-nimble vaccine industry to supply us with better-than-BioThrax countermeasures. In the last decade Washington has funded the development of at least three new bioengineered anthrax vaccines, signed up one tiny company to provide seventy-five million doses for almost $1 billion, canceled the contract for nonperformance a few years later, signed a $450 million stopgap contract for nineteen million doses of BioThrax, and initiated another round of bidding for twenty-five million doses of a new and improved vaccine. The blueprints for the four anthrax vaccines have changed corporate hands four times. As of 2011 nothing had quite worked out.

————————

AN OMNIPOTENT BUYER can easily devalue intellectual property, depress prices, deter innovation, repel new capital, promote consolidation among suppliers, and erode market resilience. With vaccines and antibiotics, that's exactly what Washington has done.

The giant drug companies that have the know-how and resources to beat the germs now operate like big defense contractors, mirror images of the insurers, regulatory agencies, and lawsuit mills that they answer to. They have lost their dynamism, flexibility, and reserve capacity. And when they do still use their assets brilliantly, getting the product to market takes a very long time and is horrendously expensive. Their products thus lag years, if not decades, behind what molecular science makes possible.

Whenever the big drug companies catch up, creating a new vaccine or antibiotic that works well against a disease, Washington grows reluctant to approve further drugs that target the same disease in some different way. After all, the disease isn't much of a problem anymore, but the new drug might be, and will undoubtedly cost more. Drug companies see this reluctance coming and dump Washington before it dumps them. Later still, when a new epidemic emerges, maybe they consider coming back.

Absent fundamental changes in public policy, we are thus condemned to relive the HIV story again and again, in an endless cycle of mass death followed by a long war against the latest mass murderer. How long, nobody knows. HIV itself is still on the march—the global HIV-positive population has stabilized at around thirty-five million only because infected children and adults are now dying about as fast as children and adults are being infected, which is to say, at a rate of about two million a year.

13

DOLLAR DOCTOR SCIENCE

CHARLES DE GAULLE once wondered how anyone could govern a nation that had 246 different kinds of cheese. Our biochemists could probably stock that many varieties on just the cholesterol shelf of the *fromagerie*. And while medicine might not need quite that many cholesterol-lowering statin drugs, it certainly needs more than one, because different statins have somewhat different effects in different bodies. The CEO of Novartis switched to Pfizer's Lipitor after finding that he responded less well to his own company's Lescol.

But for Washington's dollar doctors, our biochemical diversity is a disaster. In the short run, fragmentation in the pharmacy is expensive. Every new drug comes with a new patent that keeps prices high while it lasts. The steady arrival of new drugs offers doctors and patients a steady stream of new opportunities and reasons to go their separate, unruly, and expensive ways.

For poor countries, the simple solution is to let pirates manufacture knock-offs. Wealthy countries instead create a buyer's monopoly that gives them the bargaining power to push down prices sharply even while drugs remain protected by patents. A national pharmacy can save the government still more by stocking fewer, older, cheaper drugs—the government's favorite Brie, but not the Bleu de Termignon—and seeing to it that every patient migrates to a cheap generic the moment one becomes available.

In the United States, however, talk about the need to limit patient access to various treatments can be politically suicidal. So we have arrived instead at the "drug modernization" provisions of the 2010 Affordable Care Act and the Patient-Centered Outcomes Research Institute it established to take charge of Washington's "comparative effectiveness research" (CER). When two or more drugs are available to treat the same disorder, CER trials will pick the winner and banish the losers by refusing to pay for them. Professional medical organizations have been developing "best practice" guidelines of this sort for years; the dollar doctors—or at least the most ambitious in their current ranks—intend to develop their own and turn them into mandates. The former director of President Obama's Office of Management and Budget, a vigorous proponent of the CER trials, says they must be followed with "aggressive promulgation of standards and changes in financial and other incentives" to ensure that doctors toe the line.

In sum, Washington will gladly pay for the best medicine that the best scientists—Washington's—can find. There will be no rationing, no penny-pinching refusal to provide what patients need and what the officially best medicine can deliver. Just don't expect Washington to pay for anything less than the Washington-certified best.

––––––––

To BEGIN WITH, Washington will finish shutting down the most important sources of information that might prompt doctors and patients to stray from Washington's chosen paths. The FDA already has a chokehold on what doctors and patients may learn from the drug company, the private entity that has the most resources and clearest economic incentive to acquire and analyze information about how its drug performs in many different patients. After the drug is licensed, the company must immediately share with Washington anything new it may learn about bad side effects. As for good news, if the company learns that the drug benefits patients in ways not covered by the license—Viagra, say, is also good for some premature babies suffering from breathing problems—it may seek permission to conduct new clinical trials and amend the license. Either way, Washington strictly controls and limits what the company tells doctors or patients. The FDA must approve any changes in the label or package inserts, and the agency also spells out what the company may say in print, in broadcasts, in telemarketing pitches, and on the Web. Its rules address toll-free numbers, Internet supplements, paper brochures, and mail inquiries. They also list the government-approved medical journals whose contents may be mailed to doctors by the drug company.

Making sure that doctors and patients follow Washington's scripts will require more work. Under the authority of the Affordable Care Act, Washington will now make sure that doctors use sniffers and other diagnostic tools comparatively effectively—CAT scans, MRIs, X-rays, sonograms, genetic screening, the whole lot. A single committee, the U.S. Preventive Services Task Force, will decide which preventive diagnostic tests must be fully covered for which classes of patients by all private insurance policies, and which are deemed unjustified on either scientific or economic grounds. Insurers must cover some viral infection and cancer screens but not others. What insurers may charge is regulated, too, so more money spent complying with screening mandates will inevitably mean less spent on disfavored diagnoses and treatments to address what they reveal.

Further details of how doctors will be kept in line will emerge as the aggressive standards and financial incentives crystallize. Supervision will occur over a new national network (overseen by a national coordinator for health information technology) that funnels medical data from providers and patients to Washington's scientists, statisticians, and paymasters. The closest digital model for this network is the system that links millions of employers, accountants, and bookkeepers to the IRS, whose computers review returns for errors, flag some for audits, churn out the form letters that follow, and notify higher authorities that a taxpayer seems to have gone astray and needs a whiff of aggressive incentive to encourage compliance. Washington assures us that privacy will be protected.

The patient's privacy rights will be bought, however, when the care is paid for by Washington itself, or by any private insurance company operating under Washington's rules. Dr. Griswold must surrender control of her records if she wants Washington to pay for the time she spends prescribing the birth control pill or ordering a mammogram, and she will be reimbursed only for services that follow the approved script. Where that leaves the third clause of her Hippocratic oath—"I will prescribe regimens for the good of my patients according to my ability and my judgment"—is her problem, not Washington's. Scripted diagnostic tests linked to scripted prescription protocols will determine what her patient really needs. If Dr. Griswold believes that the script is dead wrong for this particular patient, too bad. If she doesn't care to sell her professional judgment to Washington, she must either give away her services or send the patient elsewhere. Or she can do as Washington instructs, in which case, as one doctor has observed, Hippocrates gives way to veterinarian ethics—the sick dog's treatment is determined by the master's willingness to pay.

Washington believes that it has also worked out how to deal with the First Amendment, though several courts have concluded otherwise, and the issue will almost certainly require Supreme Court resolution sometime soon. In 1998, the FDA grudgingly allowed drug companies to start informing doctors about promising off-label uses of a licensed drug—so long as the informing occurred only by way of textbooks, professional seminars, and peer-reviewed professional journals and didn't begin until the drug company had initiated the process of amending the drug's license. A federal judge struck down these conditions as unconstitutional. Federal law allows doctors to prescribe off label, the judge reasoned, and Americans—including those who speak for corporate America—"do not need the government's permission to engage in truthful, nonmisleading speech about lawful activity."

The government filed an appeal. As the appellate court tartly noted, its brief was "somewhat unclear," but no matter—the stage was set "to consider a difficult constitutional question of considerable practical importance." When the government's lawyer stood up to argue his case, questions from the bench soon made it clear that the FDA was about to lose big—at which point the lawyer declared that it had all been a misunderstanding. The challenged law didn't prohibit or threaten to punish anything at all—it just created a safe harbor. Drug companies that engaged in off-label chatter might be prosecuted for violating some other law, but that possibility would have to be addressed in some other lawsuit.

Washington's solution, we now know, is to characterize unscripted chatter as economic fraud. Enacted to help protect Washington from fraud by military contractors, the federal False Claims Act now serves as Washington's most powerful silencer of drug companies. To promote off-label prescriptions, the government maintains, is to promote the submission of false claims in connection with the care of the poor, the elderly, soldiers, veterans, federal employees, and federal prisoners. Trial lawyers—now called "recovery audit contractors"—initiate massive false-claim lawsuits on behalf of the government, and take a munificent cut when they win.

As Washington sees it, scientific truth is irrelevant in the economic fraud cases. So are the prescribing doctor's medical judgment and the patient's informed choice. The fraud lies in extracting money from Washington by saying things to others that Washington hasn't certified to be true, at a time and in a manner that Washington hasn't certified to be good. Pfizer may be sued if it breathes an unapproved word about Viagra to any neonatalogist who prescribes the drug to infants and then bills the government. Even if every word breathed has an anchor in rock-solid science. Even if, by the time Pfizer gets hauled into court, the FDA itself has reviewed the science, accepted it, and approved Baby Viagra.

The drug company may also face criminal prosecution. In late 2012 another appellate court upheld a sales representative's First Amendment right to discuss off-label uses of a drug with a doctor. A co-defendant doctor, indicted because he had made paid appearances to discuss the off-label use of the drug at events sponsored by the drug's manufacturer, had his assets frozen and his medical practice all but ruined. He pled guilty to a misdemeanor and hanged himself not long after.

In 1902, when J. H. Kelly was peddling magnetic cures, it was up to the government to prove that the junk science was junk, and Washington could expand its power only by improving its science. Today's economic frauds can involve safe-and-effective medicine and FDA-caliber scientific truth that aren't safe, effective, or true, legally speaking, because the FDA hasn't yet decreed that they are. Nothing is true until Washington has approved the script.

The expert advisory committees, doctors, researchers, and academics who help Washington script that truth thus become the unwitting allies of Washington's paymasters. Their work doesn't just help clarify and affirm the best medical science known today—indeed, because it is anchored in the statistics of crowds it does quite the opposite. It will also be used to limit the discussion of what's best, and thus limit the pursuit of better.

———

AT BEST, THE biochemically indiscriminate study of crowds leads to treatments that subdue the most common form of the disease and that can be tolerated by the most common brands of stomach, liver, kidney, heart, and immune system. Even drugs that suit the majority quite well will often fail to win the patient polls by margins big enough to persuade first the FDA's statisticians and then the Treasury's that the benefits to some outweigh the risks to others. But the advance of molecular medicine, as we have seen, hinges on doing the opposite—identifying the molecular factors that explain why the same drug has different effects in different patients, and developing enough different drugs to deal safely and effectively with all the diversity.

Until biochemists learned to distinguish estrogen-positive and estrogen-negative forms of breast cancer, for example, any comparative trial of an estrogen blocker and estrogen itself would have strongly favored the blocker, because the ER+ form of the cancer is much more common. But the crowd doctors might alternatively have concluded that neither class of drugs should be approved, because blocking estrogen would have accelerated the progress of the disease among a minority of patients—those with ER− breast cancer, we now know. Biochemists and oncologists are rapidly acquiring diagnostic tools that identify the rare but aggressive forms of other cancers that justify the use of the most potent—and therefore often toxic—drugs. Until such tools are developed, milder therapies will always look better, even if they can't help the most desperate patients.

The crowd doctors are forever playing catch-up, always lagging far behind what the best medicine can deliver. To keep crowd prescriptions up to date, the guardian of the statistics must update the databases, subdivide the crowds, and recrunch the numbers as fast as new biochemically distinct forms of the disease are identified, new drugs are developed to treat them, or new ways are found to predict how well patients can tolerate the various drugs that might be used. The crowd doctors can't start counting bodies until they have the experience of many bodies in hand. They can't start comparing the efficacy of drugs until others have spent huge sums developing two or more of them. They can't link a drug's different effects in different patients to patient-specific biomarkers until biochemists have identified the markers and doctors have tracked them in many patients.

And when the crowd doctors get something wrong, they get it wrong for the whole crowd. In an article published in early 2010, Dr. Jerome Groopman, a Harvard professor of medicine, AIDS and cancer researcher, and author of *How Doctors Think*, reviewed how Washington's "choice architects" have performed over the past decade. They have "repeatedly identified 'best practices,'" he concludes, "only to have them shown to be ineffective or even deleterious." Their mistakes have spanned the control of blood sugar levels in the critically ill, the prescription of antibiotics to emergency room patients upon arrival, and the treatment of anemic cancer patients with a growth-factor protein. Such errors have led to colitis, higher mortality rates among diabetics, and the misdiagnosis of heart failure or asthma as pneumonia. Dr. Groopman acknowledges that one standard that he himself helped develop was later found to raise the risk of strokes and heart attacks significantly. Many other standards had no discernible impact on the quality of treatment. "There is a growing awareness among researchers, including advocates of quality measures, that past efforts to standardize and broadly mandate 'best practices' were scientifically misconceived. . . . The care of patients is complex, and choices about treatments involve difficult tradeoffs. That the uncertainties can be erased by mandates from experts is a misconceived panacea."

Left to their own devices, front-line doctors lead the development of precisely targeted therapies as they diagnose, prescribe, track responses, and adjust treatments accordingly. They will be the first to see effects that were missed when the last CER study was consecrated in Washington five years ago. A different class of estrogen-inhibiting breast cancer drugs may perform better than tamoxifen in a first comparative study—but statistics supporting that view have to be discarded when researchers then discover that quite a few women lack the gene that metabolizes tamoxifen into the components that do the real work, and start prescribing the drug more narrowly. "In the case of prostate cancer," one oncologist observed in 2010, "[scientific] progress is so rapid that the use of historical data for definitive answers is not a worthy expenditure of time or money according to all of the experts I know."

The process of developing the biochemically diverse array of drugs needed to mirror the biochemical diversity of human bodies already starts badly because the FDA's own rigid, expensive trial protocols generally preclude the adaptive search for biochemical details required to arrive at a good drug-patient fit. When dollar doctors then add a second tier of crowd-based averaging, we end up with a system tailored to the generic doctor who prescribes drugs in biochemically indiscriminate ways, as directed by a generic script, written by a distant committee.

By MOVING THE drug evaluation process even further from the discipline of rock-solid, one-on-one molecular science, the dollar doctors make it even easier to slip nonmedical policies into the process. Indeed, they cannot avoid muddying factual waters with subjective opinions. The answers don't magically become scientific when inherently unscientific questions are presented to a panel of experts convened by Washington. Suggesting that Washington's pronouncements about comparative effectiveness merely affirm the best medical science and will ensure the best treatment for all is politically clever but scientifically dishonest.

The muddying is deliberate when crowd doctors discourage diagnostic screening. Washington's Preventive Services committee usually demands that diagnostic procedures prove their worth in randomized, FDA-like trials in which the patients who get the diagnosis experience better clinical outcomes later on. But the likelihood that early screening will help depends on how much medicine knows about the risk factors that launch and propel a disease, and what tools it has to neutralize them. Medicine's understanding of those risks emerges only as it acquires genetic, lifestyle, and medical data from both healthy and sick patients. The development of drugs to control risks embedded in human bodies hinges on identifying good targets, and one way to do that is to develop large databases of patient records that begin with genetic profiles and track health for decades.

For crowd doctors whose paramount objective is to give everyone the same access to the same care, it is all too easy to overlook differences that might stand in the way of one-size-fits-all medicine or magnify differences—rare side effects, for example—that tilt the scales in a desired direction. Science doesn't tell panels how to balance multiple risks against multiple benefits when both are scattered unevenly across patients. And nobody can possibly believe that government committees, unlike doctors and patients, make it all scientific when they sit in judgment on a risky drug that offers the terminally ill a chance to extend life briefly or modestly improve its quality, or that offers the genetically unlucky a chance to prevent the horror of slowly but inexorably losing all control of body or mind. When they pass judgment on imperfect treatments for a disease such as multiple sclerosis, crowd doctors can't avoid conflating science with speculation about wishes, hopes, and fears as far removed from science as the ineffable loneliness of suffering and death.

At the opposite pole, Washington must deal with drugs that are popular, cheap, and widely used by the seemingly healthy. The most revealing illustration of how easily the crowd doctors can manipulate the process that they characterize as a scientific search for the most effective medicine is found in the one area where many of them insist that the patient has a right to choose for herself.

In early 2009 a federal judge appointed by President Reagan concluded that George W. Bush's FDA had acted unconstitutionally when it denied under-eighteen teenagers over-the-counter access to the Plan B morning-after pill in response to political pressure from the White House. Not long after, an advisory committee convened by President Obama's FDA recommended the approval of Ella, a drug that prevents or terminates pregnancy up to five days after conception. "Supporters and opponents both said the decision marked the clearest evidence of a shift in the influence of political ideology at the FDA," the *Washington Post* reported. The recommendation crept its way through the agency, and by late 2011 FDA commissioner Margaret Hamburg was ready to sign on. But in what the *Post* now described as an "an unprecedented step" involving a decision that "could have saddled the Obama administration with a political target as the 2012 presidential campaign moves into full swing," Health and Human Services secretary Kathleen Sebelius overrode the decision at the last minute. An assistant FDA commissioner who had resigned in 2005 to protest the Bush FDA's action was "stunned." The Obama White House quickly denied any involvement. In April 2013, describing the administration's action as "politically motivated, scientifically unjustified, and contrary to agency precedent," the same federal judge overruled the agency again.

In the interim, Washington had notified Belmont Abbey, a small Catholic college near Charlotte, North Carolina, that in refusing to cover contraceptives, its employee health insurance plan was engaged in unlawful gender discrimination. In early 2012 this policy erupted into a national debate about whether Catholic institutions in general could be required to provide free Ella and such to nuns. The Obama administration defined contraception as "preventive" care, a decision that will force insurers to cover it without co-pays. One of the medical justifications offered in favor of this shift was that women have safer, healthier pregnancies when they can give their wombs a rest in between. But too much rest for too long can cause other problems—a child's risk of being born with or developing Down's syndrome, autism, schizophrenia, Alzheimer's, and a raft of other genetic disorders rises relentlessly as the mother and father grow older.* And as we have seen, a lifetime of resting the womb increases the risk of breast cancer—those statistics were compiled in Padua centuries before Washington got into the act.

---

* The average first-birth maternal age in the United States rose rapidly in the 1970s and has continued to rise since; the first-birth rate in women over thirty-five rose eightfold between 1970 and 2006. A 2009 statement by Britain's Royal College of Obstetricians and Gynecologists places the availability of contraceptives first in its list of the "manifold and complex" factors that account for the rising incidence of "pregnancy complications" and "adverse pregnancy outcomes," and identifies maternal age as "an emerging public health issue."

Brazen manipulation of the pill's statistics exposes the flabby unscientific underbelly of crowd medicine at its worst, because fertility isn't a disease; ovaries are supposed to ovulate. We have the pill only because so many women want to be able to turn off this one part of their normal physiology; other women pay dearly for drugs to turn it on. The pill is good medicine not because science establishes that it is, but because it enhances a private freedom that many women value—the freedom to make private choices about private matters without the crowd's permission or approval.

There is no end to how far statistical legerdemain can be pushed when analyses masquerading as science are cut loose from objectively measurable scientific criteria. If the pill makes women happier, crowd studies can also prove that happier is generally healthier. Any aspect of health that implicates autonomy and happiness cascades through the rest of life, even if it isn't quite as profoundly life-changing as pregnancy. And if popular, happy, and sexy count in drug science, Viagra, testosterone, baldness cures, and wrinkle cream can improve health, too. Stress and lonely weekends, perhaps caused by baldness and wrinkles, may also lead to more drinking and smoking, less sleep, riskier sex, and a weaker immune system—statistics suggest they do.

But what these statistical games really do is conceal the fact that such drugs are plebiscite drugs—Washington is willing to pay for them because they perform well in social and political trials.

IF AUTONOMY OR happiness can help get drugs into the crowd doctors' pharmacy, cost can help keep them out, and inevitably will. Some policy makers have already made clear that controlling costs is as important an objective as effective care. Before he became an influential health care advisor to the Obama administration, Dr. Ezekiel Emanuel argued that medical students must be trained "to provide socially sustainable, cost-effective care." Doctors must consider social costs as well as individual benefits when deciding how to treat the individual patient. Insurance companies should cover only the treatments that work for most patients. Washington's paymasters should follow the example set by Britain's National Institute for Health and Clinical Excellence (NICE), which expressly ties the adoption of new medications to how much Britain can afford to pay for every extra month of life that a treatment may provide. An illustration of how the NICE perspective changes things came in March 2011, when the FDA approved a new lupus drug—the first in fifty years. Six months later, NICE concluded that the drug does not offer "good value for money, compared to standard care" and therefore "could not be considered a good use of [National Health Service] resources."

Such views are too politically delicate to be candidly aired after their proponents arrive in Washington, but the government's dollar doctors can easily hide money policies under the lipstick and powder of crowd science. Paralysis by analysis will often suffice, and is politically easy—all it takes is the dogged insistence that we don't yet know enough about the new drug. That saves money, in the short term, because it always favors continued use of the older drug while the clock runs down on the new drug's patent. Washington almost always knows more about older drugs, because they have had years of additional vetting in the market. Moreover, the first drug developed need only prove that it's better than the disease. The second drug will often have to prove that it's better than the first, and exposing small differences requires larger, longer, more expensive trials.

THE LONG TERM is quite another matter, but no one doubts that Washington can save money in the short term by curtailing access to the diagnostic tests, medical procedures, or drugs that it pays for. But if that is the objective, it should be publicly acknowledged, as it is in Britain, and backed by cold-blooded cost-benefit calculations that weigh the drug's price against the value that Washington places on average extra months of average life and other crowd-averaged medical benefits. What Washington should not be allowed to get away with is untenable assertions that the comparative effectiveness studies and scripted protocols overseen by dollar doctors will curtail access only to inferior forms of medicine that nobody needs.

Rigorous adherence to checklists, standardized routines, and scripted protocols plays an essential role in flying planes, operating nuclear power plants, and making sure that the surgeon scrubs up and amputates the right leg, not the left. But the crowd doctors' body counts are useful, if ever, only where procedures, protocols, and treatments are so mature, static, and uniformly safe and effective that they are unlikely to improve going forward. Few areas of molecular medicine can be characterized that way. Going forward, the best molecular medicine will be choreographed by complex algorithms and analytical engines powered by continuously evolving compilations of molecular and clinical data.

Crowd doctors save money by throttling the slow, messy, fragmented, and therefore expensive process of searching for drug regimens tailored to the complex biochemical diversity that shapes our health. What patients get instead is treatment for the generic patient who dwells only in Washington's computers. That sets the stage for the government to buy each drug from the supplier, or perhaps the small group of suppliers, that offers it at the lowest price. Because there are such large economies of scale in this industry, especially when doing business in the labyrinthine corridors of government, smaller suppliers will gradually drop out of the market. Every disorder will come to depend on a cluster of large providers, or perhaps on just one.

That will take us to the threshold of what many critics of Pharma say we need. Pharma, one argues, should be viewed as a public utility and regulated accordingly. We should shorten and narrow the scope of drug patents and tighten the already tight limits on Pharma's dealings with clinical researchers, doctors, and patients. We should require that the industry open its accounts to the public and demand that drug prices be both "reasonable" and "as uniform as possible for all purchasers."

We already know where that policy ends. As we saw in the previous chapter, the vaccine sector of the market started heading down that road decades ago, and reached the finish line in the mid-1990s.

14

THE RISING COST OF HELPLESS CARE

THE REVOLUTION IN molecular medicine is still young, and many now see it as just one more thing to blame for the relentlessly rising cost of health care. The "major contributor" to rapidly rising health care costs is "the constant introduction of new medical technologies, including new drugs, devices, and procedures," says Dr. Ezekiel Emanuel, President Obama's former health care advisor. When dressed up in econometric analyses, such views bring to mind the great but irascible physicist Wolfgang Pauli, who dismissed a paper he had been sent to review as not even good enough to be wrong.

To begin with, "health care" is too elastic a term to have a measurable cost. Paymasters don't agree on what "care" that term should include, nor on which types of cost should be tracked. Much of what they do include is palliative care that was once provided in the home, mostly by women whose tireless labor was (and still is) conveniently viewed as free. To this day, on the government's books, lying at home in bed for a week because no drug can beat the fever costs nothing. Same with the unpaid time your main squeeze spends looking after sick you. And the paymasters discern cost savings in policies that few ordinary people would view in quite that light. Contraception—or, failing that, a Plan B pill or an abortion—provides the cheapest pediatrics. The biggest cost of all—the cost of failure—isn't a cost, it's a saving. Heavy smokers can be quite cheap, especially if they're also morbidly obese—massive heart attacks are quick, cheap killers. Assisted suicides are a real bargain. Quick failure looks cheap when the only alternative is expensive but helpless care.

"The cost of health care" also conflates two fundamentally different types of costs—the cost of hands-on services, which keeps rising, and the cost of molecular medicine, which, over the long term, keeps falling. And the cheap molecular antidotes eventually substitute for many of the expensive hands. The cost of smallpox care has hit zero. The cost of heart attack care has dropped from "nothing for a while, then a fortune for heart surgery" to "$20 a month for Lipitor." The economic picture is further complicated, however, by the fact that the doctors grappling with intractable diseases play a large and essential role in getting molecular medicine to that point, and as discussed further in Chapter 15, any useful economic analysis must take that into account as well.

Paymasters concerned with the economic bottom line do not have the luxury of dwelling on such details. To assess how we're doing and where things are headed they have to define "normal health" and focus on easily tracked, aggregate indicators of success or failure. To control spending, they have to make crowd-averaged calls about the types of care that should be provided. When they address "access to health care" they focus mainly on access to doctors and hospital beds; much of the quality-of-care focus, as we have seen, is aimed at curtailing what can be tagged as unnecessary or ineffective. But for rich and poor alike, the access that very often determines how the care ends is access to the right drugs.

Almost everything that the paymasters think they know looks quite different when viewed in light of what molecular science has revealed about the diverse genetic and lifestyle determinants of human health—which correlate quite strongly with gender, race, ethnicity, and other demographic factors—and what molecular medicine is already able to treat, or might well be within the next decade or two. But preoccupied as they are with current accounts, the paymasters have little choice but to focus mainly on yesterday's hands-on medicine. Washington has dug itself into an economic pit. It has promised—but can't afford—to pay for the expensive doctors and beds that keep getting more expensive but are unable to provide effective care without the right drugs in hand. Having made that promise, Washington can't afford a huge, additional up-front investment in the molecular medicine that ends up both effective and cheap.

———

BECAUSE THOSE INVESTMENTS are indeed huge, bankrupt paymasters have every incentive to emphasize how much new drugs cost, downplay how much they might deliver, and take aggressive steps to pay less for them. Over the long term, however, these policies will ensure that health care costs rise rather than fall. Experience establishes beyond doubt that molecular medicine supplies by far the most cost-effective health care, and few can doubt that it now has the power to deliver more, faster, than it has ever delivered before.

In rich countries, average life expectancy rose by thirty to forty years over the course of the twentieth century. Many experts attribute much of the increase to the sharp drop in infant mortality, which we can't repeat, and therefore doubt that life expectancy will increase by much more than a decade or so in this century. But as the late Nobel economist Robert Fogel pointed out, experts have a long history of predicting that the longest possible life expectancy is no more than a decade or so beyond the average prevailing when they make their prediction. When we look at specific disorders, a quite different picture emerges. We can already slow down many diseases that until recently were widely viewed as forever incurable—they were just too complex, too deeply rooted in our own chemistry. We still develop chronic diseases as we age, but we now develop them a decade later, they are less severe, and we control them much better with drugs or surgery.

More fundamentally, what determines a nation's average life expectancy is a broad mix of genes and lifestyles, which vary widely across different demographic groups. Even when health is improving in every group, nationally averaged health can stand still or even decline when the mix changes—when, for example, the improvements are smallest in the fastest-growing groups. When statisticians focus on the highest life expectancy achieved within demographic groups, they consistently see over the past 160 years a steady rise in life expectancies, at a rate of slightly over two years per decade, among both men and women. Better sanitation, healthier diets, and other factors associated with rising wealth account for a significant fraction of these gains. But medicine accounts for the rest.

That brings us to aging, the affliction so seemingly inevitable and inescapable that we have always viewed it as normal, not a disease at all. Washington can't afford a broad medical assault on aging, but Wall Street could, and as a practical matter, much of that assault will address the chronic diseases of the elderly. Any significant advance in this area could play a large role in rescuing the bankrupt paymasters, though they will be the last to admit it.

Much of our public health care budget is devoted to care of the elderly. As Fogel outlined, that care can be divided into two stages. Costs rise quite slowly as a patient ages, then rise sharply in the last two years of life—two years that account for 40 percent of all Medicare expenditures. This spending pattern has held steady for the last two decades.

During the first stage, chronic diseases treated with a mix of drugs and surgery account for most of the spending. Costs could be sharply lowered by drugs that delay the onset of chronic conditions, reduce their severity, and substitute for some significant fraction of the surgery. Drugs that prevent the development of major chronic diseases entirely would have a dramatic impact. According to some estimates, health care costs will stabilize at the current level (about 15 percent of GDP) if the prevalence of chronic diseases drops about 1.5 percent a year. In the 1980s, the drop rate was about 1.2 percent. In the 1990s it reached about 2 percent.

Reducing medical costs during the expensive last two years of life is a much tougher challenge. However little they care to discuss the truth in public, this is where government paymasters can—in a rational if cold-blooded way—save taxpayers a great deal of money in the short run without doing much immediate harm to the average us. After all, these are, by definition, the years during which medicine fails, and spending so lavishly on failure makes no sense.

But when does the two-year countdown begin? Washington knows all the averages, but for the individual patient, medicine can answer that question only after the fact. Is failure inevitable? Whether a countdown should begin at all depends on what kinds of treatments are available and how well they are likely to perform in each individual patient. Doctors are quite good at deciding when their own hands-on care has reached its limits, but the limits of molecular medicine are nowhere in sight.

Exploring those limits won't be cheap—the late stages of life-threatening diseases present the most biochemically complex challenges. Effective treatments will require complex drug regimens that vary widely from patient to patient. The first drug developed—say, to deal with Alzheimer's—will often deliver only a small benefit to a small fraction of the patients who are battling what looks to clinicians like the same intractable disorder. The molecule-by-molecule process that assembles the biochemical arsenal needed to help most patients will take much longer. But that is the process that transforms the untreatable and expensive into the treatable and (eventually) cheap—as it did with AIDS and HIV, and is now doing with many cancers.

The paymasters also object to diagnostic tools that flag genetic and other problems too far in advance, often on the ground that something else will probably kill the patient sooner. (Smokers use similar might-be-run-over-by-a-truck-tomorrow logic to rationalize why they don't quit.) But medicine finds out how diseases might be treated at different stages by developing drugs that attempt to intervene at every stage, and testing them in many patients.

Researchers are now pursuing the possibility that drugs could slow down what both the FDA and the paymasters refuse to view as a potentially treatable disorder—the aging process itself. There is no normal here, either. Progeria is a rare genetic disorder whose victims age so fast they die in their teens. Longevity clearly has a strong genetic component—it runs in families, and women live longer than men. A wide range of genes and proteins make some people much less likely than others to develop many degenerative diseases.

These genes suggest that Methuselah drugs are at least biochemically possible, and they give us some inkling of what they might look like. Engineers are developing nano-scale technologies to insert drugs precisely into the deepest recesses of our bodies. Biochemists are now able to construct inert, molecular-scale pillboxes that will pop open their lids only when they reach intended molecular targets, or even assemble the potent drug from inert building blocks after they reach a target cell. Using viruses and retroviruses as their syringes, molecule doctors can now insert drug factories into cells or patch them into the core genetic code of the human body. "If we know which genes control longevity," one British researcher observes, "then we can find out what proteins they make and then target them with drugs. . . . We need to reclassify [aging] as a disease rather than as a benign, natural process."

The standard response to all arguments that rising costs reflect the rising quality of care is that the never-ending growth of the fraction of our GDP devoted to health care must end regardless, because if it doesn't it will swallow our entire economy. But drugs that start expensive end up cheap. And what makes long-term economic sense for the nation as a whole isn't determined by what makes short-term economic sense to paymasters in Washington.

As Fogel pointed out, most of the growth in health care spending correlates with rising GDP—an extra $1 in family income leads to a $1.60 increase in spending on health. We demand more medicine than our grandparents did partly because we are wealthier and partly because today's medicine can deliver so much more of what we value above all else. Nobody argues that we must be spending too much on digital technology because our grandparents spent so much less.

Perhaps individuals and families, unlike paymasters, also recognize that good medicine makes them richer, not poorer—and thus able to afford even better medicine. Many people today remain economically productive long past the age at which their parents or grandparents died or lost their ability to work. The late British economist Angus Maddison estimated that humanity owes as much as half of the eightfold increase in global per capita income over the last two centuries to improvements in health and life expectancy. There is, as well, the obvious fact that ordinary people value more years of healthy time here on earth even when they are spent on the golf course or at the bridge club. According to one 2006 estimate, Americans valued the roughly four-year extension in life expectancy that improved cancer treatments delivered between 1988 and 2000 at about $2 trillion. Each additional 1 percent reduction in cancer mortality would be worth about $500 billion; a complete cure, about *$50 trillion*. We don't have to choose between health and wealth—even when they start expensive, the molecular cures that work end up making us wealthier, too.

The paymaster perspective on health care, however, takes us as far as we can possibly get from the molecular level where the real cures happen, and doesn't even attempt to consider how much those cures are worth to individual patients. Here we reach the pinnacle of government audacity, the conceit that is literally fatal. It is the belief that Washington can determine when your time has come to die. Certainly not by convening a "death panel" for you personally—that would never do—but by convening a science panel that will declare, as it decides whether to license drugs or pay for them, that the crowd that you have now joined has received its sentence of death from the Almighty, and further treatment is futile.

By deciding when it's too early to worry or too late to help, the paymasters can adjust the line between healthy and sick to ignore slow, microscopic death sentences and certify death itself as just another aspect of normal health. The individual patient rarely makes that mistake, and this undoubtedly explains why the paymasters consistently fail to exercise their power aggressively enough to reduce or even stabilize spending—voters are or expect to be patients, too, and they just won't stand for it. The individual craves eternal youth, with benefits, and knows such cravings are altogether normal and healthy. Normal people don't wish to be normal on the pitcher's mound, in the mirror, or in bed—they wish to be above average now and forever. Given the choice, many would opt to start fighting genetic or lifestyle fate sooner rather than later. Many prefer not to succumb to leukemia tomorrow if living another year or two is an option. The twin impulses to stay alive and create more life lie at the existential core of life, as deeply embedded in people as in viruses. The biologically normal view of health care reflects an insatiable appetite for whatever can deliver more years of active existence on our small jewel of a planet.

Untreatable diseases are expensive because they don't always kill fast enough to make death cheap, and the very sick are often willing to spend as much as they have—or as anyone else will give them—to stretch life out a bit longer. "All my possessions for one moment of time," Queen Elizabeth is supposed to have said as she lay dying in 1603. Lots of ordinary people feel the same way. The only financial crisis we face in health care is the one made in Washington—a crisis of open-ended promises irresponsibly made by cynical politicians who are forever willing to promise what they cannot deliver.

———

THE PAYMASTERS AND their defenders often respond that the problem lies not in the promises but in the inefficiency of the delivery system. "Cutting waste" is, in the abstract, always popular, and they have conclusively established, to their own satisfaction, that care improves and costs fall when national governments take charge of all the spending.

Typical of countless others in the field, one 2006 study—based on a telephone survey—concludes that Canadians are healthier than Americans because they manage access to health care from the top. "We pay almost twice what Canada does for care," one of the authors declared, "yet Canadians are healthier, and live two to three years longer." "This finding," according to a second author of the study, "is a terrible indictment of the U.S. healthcare system. Universal coverage under a national health insurance system is key to improving health."

A critic of the study pointed out that Americans are, on average, fatter than Canadians. "This speaks more to genetics, diet, exercise and culture than to the accessibility or inaccessibility of health services. The remedy for obese Americans will be found in less fast food and more gym memberships." But Canadians smoke more, responded the authors. All things considered, there's "not an iota of evidence that genetics or culture" account for why Canadians live two years longer, so the credit must go to Ottawa. But with both genes and culture out of the picture, we are left to wonder what accounts for Canada's different diet, smoking, and politics. Could it be hockey?

Perhaps not.* Until quite recently, America smoked more than any other country in the developed world; we also currently have far higher obesity rates. Adjust the statistics to exclude obesity and one or two other lifestyle factors, and it looks like Americans have the highest life expectancy in the world, though it's impossible to say for sure, because obesity is a medically fuzzy term and its links to so many problems, while clear, are hard to translate into average changes in life expectancy. We do know that U.S. obesity rates have leveled off in the last decade, but at different levels in men and women, and at sharply different levels in different demographic groups.

Infant mortality statistics often supply a second count in crowd doctors' indictment of health care not managed by crowd doctors. But according to Washington's own statisticians, "the primary reason for the United States' higher infant mortality rate when compared with Europe is the United States' much higher percentage of preterm births." Here, too, the rates are strongly affected by lifestyle choices and vary sharply across racial, ethnic, and other demographic lines. When the statistics are sliced more finely to reflect demographic heterogeneity, U.S. health care is doing at least as well as Canada's and Britain's, and in many respects better. When the statisticians focus specifically on premature and low-birth-weight babies, for example, they find that those born in the United States are the most likely

---

*For a trenchant discussion of the good, bad, and ugly statistics of life expectancy and infant mortality and what they reveal about the quality of health care in different countries, see Chapters 2 and 3 of Scott Atlas' *In Excellent Health: Setting the Record Straight on America's Health Care* (Washington, DC: Hoover Institution Press, 2012).

to survive. And the U.S. numbers are probably even better than they look. Most neonatal deaths occur very soon after birth, and there is reason to believe that in countries where doctors and hospitals are accountable to government bookkeepers rather than to patients, neonatal deaths are often misreported as "fetal demise" (stillbirths).

Finally, note that Swedish-Minnesotan-Americans are much healthier than the average unhyphenated American, and the real science is now centered on thousands of hyphens that slice much more finely than that. Might a different hyphen mix—rather than Ottawa's scientific management—explain Canada's cheap longevity? The most certain fact in modern medicine is that out-of-whack human chemistry is by far the biggest killer in affluent societies, and that molecular problems are caused mainly by genes and by lifestyles shaped by some mix of genes and culture.

––––––––

HOWEVER LITTLE THEY care to dwell on it, the champions of socialized medicine know this full well. They see wide gaps in infant mortality and life expectancy across different ethnic and demographic groups inside their own countries. And the European Union remains, by and large, a group of discrete, insular nations, the healthiest of which would never agree to integrate their health care systems and costs with the rest. A union that has trouble maintaining a single currency isn't about to embrace single-provider care of its biochemistry.

In the United States, Asian American women have a life expectancy of almost eighty-seven years; African American men, sixty-nine years. We have these facts on the authority of "Eight Americas," a 2006 study by number crunchers at Harvard's School of Public Health. Women in Stearns County, Minnesota, live about twenty-two years longer than men in southwest South Dakota and thirty-three years longer than Native American men in six of South Dakota's counties. The gap between the highest and lowest life expectancies for U.S. race-sex-county combinations is over thirty-five years. Some race-sex-county groups typically die in their nineties, others in their fifties. Some are healthier than the norm in Iceland, Europe, or Japan, others sicker than Nicaragua or Uzbekistan. As the authors of "Eight Americas" write, "The observed disparities in life expectancy cannot be explained by race, income, or basic health care access and utilization alone." Low-income whites die four years sooner in Appalachia and the Mississippi Valley than they do farther north. The healthiest whites are low-income residents of the rural Northern Plains states. In the West, American Indians who remain on the reservation die much sooner than whites.

What accounts for these cavernous differences? Harvard names six leading "risk factors" for the population as a whole—alcohol use, tobacco

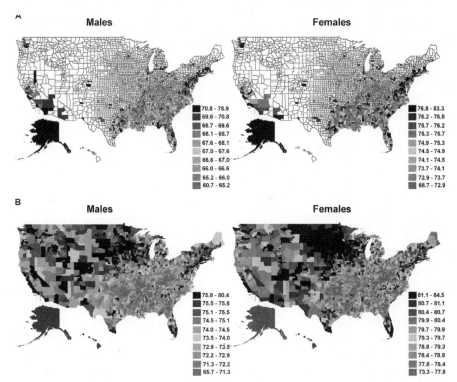

FIGURE 14.1 County life expectancies by race, averaged for 1997–2001. (A) Life expectancy at birth for black males and females. (B) Life expectancy at birth for white males and females.
Source: Christopher J. L. Murray et al., "Eight Americas: Investigating Mortality Disparities Across Races, Counties, and Race—Counties in the United States," *PLoS Medicine* 3, 9 (September 2006): e260.

use, obesity, high blood pressure, elevated cholesterol, and elevated blood glucose—and reports that these factors correlate strongly with the life expectancy differences across its "eight Americas." One of the study's authors ventured to suggest that where you live may point, in turn, to ancestry, diet, exercise, and occupation.

This is timid hash, and there are only two ways out of it. Harvard retreats into fog: we're beset by "socioeconomic," "cultural," and other "distal" causes of disease. The alternative is to disassemble culture, lifestyle, gender, and ancestry into components as small, specific, and measurable as cholesterol. At the far end of this politically treacherous trail, the distal fog melts away in the harsh practical light of molecular chemistry, and the drugs that we do—or don't—have to control it.

WHICH BRINGS US finally to common decency and economic justice. Much of the rationale for government-controlled health care is that it is the only way to address the unfairness of a world in which some people can afford good health care while others can't. But life itself is unfair, and while others have suspected as much before, biochemists can now prove it. And that unfairness gets replicated, unfortunately, in the drugs we develop—or fail to develop.

One lung cancer drug targets a receptor that occurs disproportionately in nonsmoking women of Asian ancestry, another the 4 percent of lung cancers that occur most often in younger nonsmokers. Most other lung cancers remain untreatable. About one in five patients with advanced melanoma respond stunningly well to two new monoclonal antibodies. When one form of colon cancer spreads to your liver, you need Pfizer's Camptosar—unless a flawed UGT1A gene means you lack an enzyme to purge the drug from your body before it accumulates to toxic levels. Or try aspirin—it seems to improve the survival prospects significantly in the 15 to 20 percent of patients who carry the PIK3CA genetic mutation. Two recently licensed hepatitis C drugs are only half as effective in white patients as they are in black patients. Codeine, the widely used painkiller, doesn't work in people who lack the gene that produces a liver enzyme that converts the drug into morphine (6 to 10 percent of Caucasians, 3 to 6 percent of Mexican Americans, 2 to 5 percent of African Americans, and about 1 percent of Asian Americans), and it works dangerously well in people who produce a lot of the enzyme (1 percent of Finns and Danes, about 4 percent of North American Caucasians, about 10 percent of Greeks and Portuguese, 20 percent of Saudi Arabians, and almost 30 percent of Ethiopians)—plus on top of that, more than a dozen other drugs, including a number that are widely used, inhibit the enzyme and thus dull codeine's effects.

Genetically discriminatory drugs reflect biochemical reality, unfair though it is. The performance of quite a number of these drugs is determined by genes that track gender, race, or ethnicity; their FDA licenses affirm truths we prefer not to acknowledge, and approve conduct that is illegal in every other sphere of commerce and public life. These drugs are also terrible news for anyone determined to pull people together, pool medicine's costs, and equalize its benefits. The doctors of equity promise universal access to the Mayo Clinic, where the real doctors now prescribe discriminatory cures and card your genes at the door.

By funding research Washington can help address the legitimate concern that market forces alone will favor the development of drugs directed at diseases that afflict the affluent and politically influential—though the affluent and influential are usually good at influencing Washington's spending priorities, too. Washington can also subsidize access to drugs, however they are developed. Beyond that, Washington can't help. No privacy-protecting, discrimination-banning law, no commitment to universal care, and no promise that someone else will pay will ensure that a drug that helps others will help you, too. Prepare for a long battle between radically new medical science and a senescent, antiscientific vision of how diseases are cured and what the health care system ought somehow to deliver at a price that Washington deems affordable.

There is no middle ground. We will have to choose—between the medicine that learns to deal, molecule by molecule, with unfair, unequal, discriminatory, fragmented, biochemical reality, and the medicine favored by crowd doctors, who cling to the view that if they scrutinize, track, certify, and choreograph things just right, they can deliver better medicine to all from afar.

Happily, however, that choice is not to be confused with the all-too-familiar, reflexive assertion that we have to choose between affordability, on one hand, and the pursuit of universal service, on the other. The cost of hands-on care dwarfs the cost of drugs and keeps rising, while the technologies that we use to design and mass-produce drugs are getting more powerful and cheaper by the day. And because so much of the cost and value of drugs consists of know-how, it is not only possible but economically efficient all around—better for both drug companies and patients—to price drugs in ways that ensure their very broad distribution.

The debate here is—or should be—about capturing the enormous economies of molecular medicine. The economic issues are subtle and complex—they always are in information markets. But because most of the value of molecular medicine does indeed consist of information, drugs can, over the long term, deliver not only the best but also the cheapest medicine, as indiscriminately and universally as nature scatters the drugs' targets across humanity.

Here again, the tidy, simplistic solutions backfire, raising costs and prices and curtailing the broad distribution of drugs. At their worst, well-intentioned but clumsy economic policies drive private capital out of some sectors of the drug market—a disastrous result at a time when so much hinges on innovation and the government, deeply in debt, is unable to make the long-term capital investments that information markets require. What we need instead are policies that encourage private capital to make the long-term investments that Washington can't afford to make, and adopt pricing strategies that spread these costs efficiently across users—which, as it happens, are strategies that also get the rich to subsidize the poor. These are the subjects of the next two chapters.

15

THE FALLING COST OF HEALTH CARE

IN HIS APRIL 2005 obituary, the *New York Times* described Maurice Hilleman as the man who "probably saved more lives than any other scientist in the 20th century." What kind of genius does it take to get *that* on your tombstone? Hilleman himself, it seems, "credited much of his success to his boyhood work with chickens." He went on to use fertilized chicken eggs to grow large quantities of bacteria and viruses that were then weakened or killed to produce vaccines. The technique was known before Hilleman arrived, but isolating and then safely breeding a pathogen requires the touch of a delicate artist. "Maurice was that artist," a leading expert in infectious diseases recalled; "no one had the green thumb of mass production that he had."

By farming eggs on a scale made possible by the resources of Merck, a huge drug company, Hilleman saved tens of millions of lives and protected countless millions more from deafness, blindness, and other permanent disabilities. No Albert Schweitzer or Florence Nightingale could claim numbers like his. And Hilleman saved lives by the carton, at grocery store prices—acres of cartons, hundreds of millions of warm eggs replicating his genius around the clock.

The cost of Hilleman's kind of medicine tracks a slope—precipice, really—that is far steeper than most of the other hills and valleys of economic life. The first pill may cost a billion dollars. Copies can then be churned out at pennies a pop. These clifflike economies rule throughout the pharmaceutical industry because so much of the first-pill cost is tied to know-how embedded in the drug, in the process for mass-producing it, and in the protocols for prescribing it—and know-how is expensive to develop but costs nothing to copy.

As the previous chapter argued, any serious discussion of health care economics must begin with getting that difference straight. It is a difference between a market for services and a market for what is mostly information—a difference between current consumption and investment in imperishable knowledge. The rest of the story centers on the front-end cost of acquiring the information that ends up so cheap.

———

WE BEGAN SYSTEMATICALLY outsourcing medically helpless health care early in the twentieth century—to thousands of nurses, orderlies, beds, and iron lungs, and to armies of attendants. Spending on such services rose fast wherever governments began paying the bills. Then, quite suddenly, the tuberculosis sanitariums, polio wards, acres of buildings, and their huge staffs all disappeared—driven out of business by the likes of Hilleman and Merck. Today's molecule doctors mass-produce their insights and expertise in sprawling complexes of glass, stainless steel, air handlers, pumps, heaters, and automated control systems—factory-like production lines kept cleaner than a hospital operating room, and operated under tighter control. And when molecular medicine gets good enough, it displaces doctors and beds completely.

The triumphant war we fought against the legacy germs has already taken us a good distance down that road. The diseases that used to kill countless children and many adults in their prime are now kept at bay by vaccines administered by nurses in doc-in-a-box clinics. Once-lethal throat and lung infections are diagnosed with a test strip available from Amazon and beaten back with a five-day regimen of antibiotics. Hepatitis C drugs replace liver transplants and other costly treatments.

The control of our own chemistry is now headed the same way. Insulin is quite potent enough to kill, but most diabetics took charge of its day-to-day use long ago, and smart blood sugar sniffers and insulin dispensers can supervise the process with minimal help from anyone. Even as the population has aged, hospital admissions for heart disease, the leading of cause of death in the United States, have fallen sharply since the late 1990s; the rising use and falling price of anticholesterol drugs played a significant role. One form of leukemia is kept under control with a daily dose of Gleevec. Nurses and pharmacists already provide much of the assistance that patients may need to monitor test results and adjust dosages of pill-a-day regimens used to treat chronic diseases. Molecular medicine thus allows the patient to move much of the care back into the home, off budget and out of sight.

Molecular medicine currently accounts for about 12 percent of the official, on-the-books health care budget, but it delivers a very large fraction of the effective treatment. Most of the medically supplied gains in health and life expectancy have come from vaccines and antibiotics, followed by drugs that tame our human chemistry before it clogs arteries, clots blood, or attacks its owner in some other way. Without drugs, today's doctors would be almost helpless; without doctors, we would nevertheless muddle our way through many medical problems by self-prescribing drugs under the guidance of labs, nurses, pharmacists, sniffers, Google, and the *Physician's Desk Reference*.

But as we muddled, we would have to remain content with yesterday's medicine. Getting molecular medicine to the autopilot stage takes a long time, and front-line doctors play a central role in getting it there. Nobody ever said that information starts out cheap. Medical information certainly doesn't.

WE OWE MUCH of our better health today to huge, forward-looking investments made by others long ago. The water lines and sewer pipes they built would last for decades. What they discovered about disease transmission still helps keep us healthy today. Their successors then developed, at great cost, the technical skill used to isolate and cultivate big clumsy bacteria and then viruses, and learned how to weaken, kill, and dismantle them while leaving enough code intact to fire up the patient's immune system (a difficult and dangerous undertaking) and how to maintain the vaccine's potency while it was being distributed (a tricky process in itself).

The antibiotic pioneers raised the information content of medicine another notch when they began searching for and finding molecules to target microbes and learned how to mass-produce them. Alexander Fleming stumbled on the antibiotic power of a fungal secretion in 1928. But when penicillin manufactured by Merck was first used in the successful treatment of a U.S. patient in 1942, that single patient used up half the total supply then available. A worldwide search led to the discovery of a very effective form of penicillin on a moldy cantaloupe found in Peoria, Illinois, in 1943. Fleming shared the 1945 Nobel Prize for Medicine with two scientists who pioneered a reliable process for its mass production.

And then suddenly, in our generation, biochemists learned how to isolate, cultivate, and manipulate molecules themselves.

The last major step in that process was taken in 1984, on a moonlit night, on a mountain road in northern California. Kary Mullis, an obscure biochemist, was toying with a problem of nucleic acid biochemistry as he drove. A sudden flash of insight impelled him to pull over and wake his girlfriend, who was asleep in the passenger seat beside him. He had solved a problem of great significance, he told her: how to replicate DNA in a glass flask. Afterward, with the wisdom of hindsight, many scientists were surprised they hadn't thought of the polymerase chain reaction (PCR) themselves. Mullis collected the 1993 Nobel Prize in Chemistry and retired to enjoy California's surfing and beach bunnies.

PCR is a delicately choreographed sequence of chemical reactions that copy genes in much the same way that life does. Seeded with a strand of DNA culled from nature or engineered in the lab, a fully automated reactor no bigger than a toaster oven (or even collapsed onto a tiny silicon chip) turns that first strand into two, and then the two into four, and goes on doubling from there on out. The process can also be tweaked to manipulate and modify the code in the genes being replicated. The worldwide market for the DNA-replicating machines and the tests they make possible is doubling in size about every three years.

Mass-produced DNA can be fed into batteries of gene sequencers—desktop-sized machines that automate a process pioneered by two scientists who won a Nobel Prize in Chemistry in 1980. Sequencers read the code embedded in a single gene, a chromosome, or an entire genome. The cost of sequencing a human genome has dropped about 50,000-fold since 2003 and continues to fall. This technology can expose all the code, in all its complex diversity, stored in the nucleic acids of humans, germs, and anything else alive.

Our ability to mass-produce genes also makes it easy to isolate and supply the raw materials that we now use to mass-produce proteins. The developers of the key tools collected their Nobel Prizes in 1978 and 1984, for the discoveries of enzymes that can cut with surgical precision across the two backbones of a double-stranded DNA molecule, stitch pieces together, and insert them back into living cells. The synthetic organism thus created then churns out copies of the protein that's coded in the gene.

Sometimes the protein itself is then used directly as a drug. When Frederick Banting walked down the diabetes ward of Toronto General Hospital in 1922, using a needle to wake children from their terminal comas to be greeted by the joyful tears of their parents, his syringe contained insulin extracted at great expense from fetal calves. Pig and cow insulin worked the (imperfect) miracles thereafter, until Genentech and Eli Lilly inserted the human insulin gene into a bacterium and brought Humulin to market in 1982.

Alternatively, our ability to mass-produce proteins now makes it relatively easy to mass-produce matching anti-proteins and use those as drugs instead. This is the monoclonal antibody production process discussed earlier, used to churn out drugs such as Herceptin. These anti-proteins can also be put to work as diagnostic tools. Copious amounts of cheap anti-protein on a dipstick can sniff out traces of almost any protein we care to detect—the hCG protein that choreographs pregnancy, for example. Sniffer molecules can now give any of us molecular vision for the price of a bottle of eyedrops. Embedded in chip-sized microlabs, they power devices that can scan for a million molecular markers in a few hours, at a cost of a few hundred dollars.

The sniffer chemicals make it possible for scientists to expose all the biochemical diversity in humanity's protein pool. The sniffers disassemble breast cancer into multiple different diseases, identify multiple receptors for drug designers to target, and tell doctors how to distinguish patients who can tolerate a drug from those who can't. Finding out why these patients differ leads to the discovery of new receptors and the development of new drugs.

And that closes the loop. The genes lead to proteins, which lead to sniffers, which are used to probe more diseases in more patients; the probing leads medical science to more genes, more proteins, and more molecular targets for more drugs and sniffers, and the cycle continues. At every stage of this process the power of the technology is rising and its cost is falling at least as fast as it is anywhere in the digital world, and in many cases faster.

The gap will widen from here on out. Digital code knows how to clone itself, but not its hardware; biochemical code is smart enough to clone both. Edward Jenner's drug factory was a ruminant quadruped, devoted primarily to producing milk, but willing to cultivate cowpox on her udder while she was at it. In the first eight decades of the twentieth century, the virus-incubating cow was joined by Hilleman's chicken eggs, and then by bacteria, live ovaries extracted from Chinese hamsters, and the live descendants of two lines of human fetal cells. Vaccines are now created by inserting a gene or two from a serious disease (such as bird flu) into a relatively benign virus (such as the one that causes the human cold) that's easy to breed. The insulin factory is now about 0.0001 inch long—a new hybrid beast roughly 99.98 percent bacterium and 0.02 percent human. The drug is code, and the drug factory is code, too. We have here a self-replicating drug factory, the chain reaction of life propelling a chain reaction of medical productivity. The code mass-produces itself.

If this were the end of the story, the end of medicine would now be in sight, as would all our economic worries about what it costs. We know how to find targets on free-floating molecules such as enzymes, or receptors on a cholera bacterium, a cancer cell, or anything else. We know how to design and mass-produce an anti-molecule that will glom on to its target when they meet face-to-face. Much of the technology that gives us these powers is now embedded in microchips or desktop-sized laboratories and controlled by microprocessors. At every step of the way, we use supercomputers and the cloud to decode molecules, link them to clinical symptoms, and design anti-molecules. It's all getting more powerful and cheaper, at an ever-accelerating pace.

———

BUT LIMITLESS SUPPLIES of a limitless variety of cheap magic bullets won't suffice—not without maps of the human and microbial terrain on which we hope to use them. As it was in Paul Ehrlich's day, drug development remains overwhelmingly a search for information, a good that, once acquired, can be shared with everyone at almost no additional cost. But much of the information is now being extracted by oncologists, immunologists, and other specialists who know where and how to look for the information that might matter. Because they are now probing the dynamics of diseases that are more complex, the front end of the molecular medicine that ends up so cheap now needs expensive doctors more than ever before, a great deal of front-end money, and patients willing to take medical risks. Moving such drugs through the FDA's conventional trial protocols has grown much more difficult and expensive. According to one estimate, getting a drug to market today takes three times as long as it did in 1950 and (in constant dollars) costs a hundred times as much.

The most valuable information we are now acquiring describes the molecular-scale workings of the patient's chemistry, and its value extends far beyond any individual drug. Much of the cost of acquiring it should be viewed as capital investment rather than current spending. What we are learning today about the architecture and dynamics of complex diseases and the side effects of potent drugs will benefit countless future patients, few of whom will pause to think how much our generation spent working out these details. Eventually the treatment of even very complex diseases will reach the autopilot stage, or something close to it. The greatest economies of molecular medicine lie in its power to squeeze the specialist doctors out of the process, one disease at a time, much as sewer pipes, vaccines, and antibiotics transferred the control of many infectious diseases to engineers, plumbers, and the doc-in-a-box at Walmart.

———

IN INFORMATION MARKETS, the real problem is making sure that costs don't fall too fast. The development of a new drug involves a complex mix of upfront costs that span the coordinated development of the drug's chemistry and the mapping of the patient-side chemistry, with the patient-side effort accounting for a large and rapidly rising share. This creates many possibilities for competitors to hitch a free ride on the coattails of pioneers.

Begin with the pioneer's curse. As we saw in Chapter 5, the development of a pioneering drug reveals a great deal about the biochemistry of the disease and the best strategy for controlling it, and thus hands competitors information that lets them rush in with slightly different drugs developed at much lower cost. The pioneer can easily end up as the only player that fails to profit from its own path-breaking work.

Figure 5.1 in Chapter 5, for example, illustrates how later drug companies attached various short tails onto the first statin molecule to create the drugs that soon eclipsed it. For another illustration, consider Genentech's decision to compete with itself. The company's Avastin suppresses a protein that spurs the development of blood vessels. In 2004, the FDA licensed it for treating colon cancer. In 2005, Genentech announced the successful results of clinical trials of Lucentis, designed to target a similar protein that also spurs blood vessel development to cause a severe form of macular degeneration, the most common cause of failing vision in older Americans. Genentech undoubtedly realized all along that Avastin might be used for macular degeneration as well. The company probably set out to develop its own Avastin copycat drug in tandem with the original in part to maintain a two-tier pricing structure. Such pricing schemes—in which a manufacturer contrives to charge different prices to different buyers for the same or very similar products—are common in certain types of market, and much criticized, but as discussed in the next chapter, they are often economically efficient. But the FDA didn't approve Lucentis until 2006, and meanwhile ophthalmologists had begun using Avastin off label to treat macular degeneration. It worked well. "Compounding pharmacies" began buying Avastin and repackaging it in tiny doses suitable for injection into eyes. Doctors then stuck with Avastin when it became clear that Lucentis would cost more.

Genentech, however, flatly refused to conduct the additional trials required to establish officially that its cancer drug was also good for failing eyes. In 2007, the company announced instead that it would no longer supply Avastin to compounding pharmacies. So in 2008, the NIH launched a comparative trial itself. As of 2011, Washington was poised to begin doing indirectly what it could have prosecuted as criminal fraud if done by Genentech—promote the off-label prescription of Avastin to patients covered by Medicare.

Then there's the problem of the pioneer's exit—drug companies sometimes abandon their drugs. The orphaning of drugs has accelerated as large insurers and government paymasters have accelerated our shift to generics sold at last-pill prices that barely cover the cost of manufacturing the drug properly. Sometimes they fail to do even that, and when generic manufacturers then cut corners, supplies of essential drugs are interrupted.

Hospitals and doctors have been reporting shortages of many important generic cancer drugs severe enough to interfere with proper treatment—as they see it, the companies that make generic drugs "have a diminishing interest in ensuring a strong supply." Two of the three generic manufacturers of a widely used anesthetic recently had to recall their product and shut down their production lines for extended periods; one abandoned the market after being assessed $500 million in punitive damages for distributing a product contaminated with a hepatitis virus. A recent wave of corporate mergers—seven manufacturers now produce about 80 percent of the generics used in the United States—has aggravated the problem. The FDA began systematically tracking interrupted drug supplies in 1999. The problem has been growing worse ever since, and sharply so in the last few years. At the end of 2011, a record 267 prescription drugs made a national pharmacists association's short-supply list.

The problem of drugs orphaned by manufacturers can probably be addressed by long-term contracts that include penalty clauses for failure to deliver. But there is also the orphan label problem. Repurposing—using an old drug to target a new disease—presents the problem in its purest form. Here, *all* of the new development centers on the patient-side chemistry. But there is no profit in searching for new information about an old drug's effects unless there is a market ahead that will cover the cost of the search. If the drug is already available as a generic, there won't be such a market. Even if it isn't, the new use may fail to boost the value of the drug enough to cover all the costs and risks involved in a market ruled by drug regulators, paymasters, and tort lawyers.

AS THE NEXT chapter will discuss, intellectual property laws determine how well, if at all, we reward private initiative that develops the molecular know-how of life, and thus they determine how much the private sector is willing to invest in developing the intellectual capital that keeps lowering the cost of health care. The governments that write those laws are major buyers of drugs, so they have strong if shortsighted incentives to keep these rights narrow and have them expire sooner.

Even while the rights last, moreover, Big Insurance frequently flexes its muscles to flatten and lower the prices a drug company would otherwise charge. Similarly, the Canadian government, acting as a buyers' cartel, may force the pioneer to choose between selling at a cut-rate price or not selling to Canadians at all. If it chooses to sell, residents of Detroit may then shop at Canadian prices by hopping on a Greyhound bus. Online shoppers can skirt the law and shop at Canadian prices anywhere.

Other drugs are just too important for their own economic good. Supplies of vaccines are especially precarious, one outspoken critic of the industry acknowledges. She also identifies the problem: in 1994, the Centers for Disease Control "capped prices of childhood vaccines the agency purchased for use in public-health centers throughout the country—the source for most children in the United States." Nevertheless, she says, drug companies ought to be willing to continue supplying vital drugs "as a social service—

and a thank-you to the public that subsidizes them so handsomely." Unfortunately, that is pretty much how things work already, with many companies that once manufactured vaccines deciding to let some other company's shareholders deliver the thank-you.

Meanwhile, the diseases are getting more complex, and the front-end cost of developing drugs to beat them therefore keeps rising. Even a perpetual, ironclad patent would not guarantee returns sufficient to cover first-pill costs—no matter how good the drug, or how desperately it may be needed. This seems obvious enough in the case of a disease that afflicts people who survive on a dollar a day—and explains why there is so little private investment in diseases that are found only in very poor countries. But the calculation has to be made every time, in every case, and it does not become easier just because people paying for the pill happen to have more money. A 2006 article in the *New England Journal of Medicine* attributed the complete absence of drugs that would prevent, rather than just alleviate, the late-stage symptoms of diseases such as Alzheimer's or osteoarthritis to a drug approval process that "makes it hard, if not impossible" to move the drug through Washington before its patent expires.

The biochemists, doctors, and patients involved in developing the molecular maps and blueprints associated with a new drug are creating value that spans a much a wider swath of biochemical space and time. This know-how, which we are developing at such high cost today, will grow increasingly valuable as it accumulates. But the interval between when the money gets spent and when it is recovered in the market keeps getting longer. And much of the time the payback isn't commensurate with the value created— the maps and blueprints of human and microbial chemistry are currently protected little, if at all, by patents and other forms of intellectual property.

Washington, in sum, is still stuck—and seems determined to stick us ever more deeply—in the current-consumption mind-set of medical economics. Require the drug company to develop all the science up front. Structure property rights in the know-how in a way that forces these high front-end costs to be recovered quickly by a handful of pioneering drug companies from the patients who use pioneering drugs while they're still under patent. Then let everyone share the benefits at last-pill prices.

These policies create an economically inefficient disconnect between what must be spent to get a new drug to market and what it's worth to patients, present and future.

16

REASSEMBLING THE PIECES: PART 3

MERCK SPENDS A billion dollars getting to the point where it can mass-produce the genius of a Maurice Hilleman at pennies a shot. A still higher cliff separates the vaccine's value to its recipient from that mass-production cost. The diseases eradicated by Hilleman's green thumb would otherwise have cost humanity countless hundreds of billions of dollars in lost productivity, disability, premature death, and time spent attending to the sick. In circumstances like these, how is Merck to set an appropriate price?

Economists have established—as rigorously as things ever get established by the dismal science—that there is no efficient price, no "right" price, for the main constituent of every drug: know-how. Any pricing scheme is, from one perspective or another, inefficient, unreasonable, or worse. Pharma's critics demand grocery-store prices pegged to the cost of the last pill, not the first. And that is indeed the economically efficient and socially desirable price to set—after a Hilleman has worked his magic. But if we peg prices to last-pill costs, we will not get another $2 trillion or so of private capital funding future Hillemans.

The best solution, if you can pull it off, is to charge both more and less at the same time. The original patents for eflornithine, a drug developed in the 1970s, covered two intended uses: inducing abortion and reducing the size of an enlarged prostate. The drug didn't work for either purpose—but it did, unexpectedly, provide a better cure for something else: sleeping sickness. Endemic in many parts of Africa, sleeping sickness is the second-deadliest parasitic disease on the planet. (Malaria is first.) Treated with eflornithine, the near-dead sleepers arise, take up their pallets, and walk. But they are too poor to pay on the way out. In 1999, the manufacturer stopped producing the drug. It licensed the formula to the World Health Organization, but the WHO was unable to find a company that was willing to manufacture it. Sub-Saharan Africa can't cover even the cost of setting up facilities to manufacture Western drugs to Western standards. The really poor can't even afford the last-pill cost.

Then another company discovered that eflornithine impedes the growth of facial hair, and began marketing it as a beauty cream called Vaniqa. The company that still controlled the rights to eflornithine now had the incentive to continue manufacturing it, and it decided to donate a five-year supply to the WHO and fund associated research, surveillance, and training for health care workers—a gift worth about $25 million. Yes, just as Pharma's indignant critics charge, the rich get Viagra while the poor succumb to malaria. But the poor now have a better way to beat sleeping sickness.

As told by the *New York Times* journalist Tina Rosenberg, the Vaniqa story is "a scandalous illustration of the politics of neglected diseases—and of how much wealthy people drive the global medical market." Pharma, she acidly concludes, values American complexions, not African lives. Well, not exactly. Shareholders paid for the drug's development; complexions paid for the factory. Many small drug companies, government labs, and academics pursue drugs that we really need. Organizations such as the WHO were keenly interested in getting eflornithine to Africa. But none could get it there, not for love or money. That required Vaniqa.

The real scandal is how difficult we make it for drug markets to do much more of the same, and do it closer to home. Under current rules there are many diseases that the money won't pursue at all. To get private money fully engaged in saving our lives, we need intellectual property rights that support pricing schemes that spread revenues efficiently across biochemical time (both early users and later users) and across biochemical space (facial hair and sleeping sickness). Properly crafted, these rights will lower costs, lower prices, expand and accelerate the development of invaluable drug science, accelerate the introduction of important new drugs, and accelerate their broad distribution to both rich and poor.

ECONOMISTS CALL IT "price discrimination." It sounds nasty and infuriates Pharma's critics. It shouldn't. Charging both more and less at the same time is how markets grope their way toward the right price when there is no right at the end of the tunnel. The best scheme all around, for sellers and buyers alike, is a wide range of wealth-adjusted (or, technically speaking, "demand-elasticity-adjusted") prices. Complicated in theory, it's often an utter mess in practice—as we learn firsthand whenever we fly. Business travelers get soaked, college students board at a price only slightly above the cost of the extra fuel required to carry the extra weight, and the jumble of prices in between drives the rest of us nuts. But the planes are packed full, and that pushes the average ticket price way down. The rich fly, and the much less rich fly, too. Any other pricing scheme that covers both the cost of the aircraft and the cost of flying it won't fill every seat, and will therefore raise the average price that must be charged.

With drugs, ideally the manufacturer will adjust prices to mirror what different communities, clinics, and patients are willing to pay, and it will keep adjusting prices as the market evolves. All big U.S. drug companies take things a step further, with programs designed to ensure that patients who can't afford to pay at all fly free. This is the wonder of know-how economics. Doctors and hospitals provide cut-rate and free services, too, but they can only go so far: when the New York matron pays her plastic surgeon, he doesn't throw in a free resurrection for the afflicted in Ouagadougou. Drug companies can and do.

Done just right, price discrimination strikes the best balance that can be struck between the need to encourage private investment in drug development and the benefits of supplying drugs at the lowest possible price to the patients who need them. Yes, this approach to pricing allows drug companies to earn more money and earn it faster. But it also gets them interested in patients, diseases, and other aspects of biochemical science that they would otherwise ignore. And it also gets the drug to many more patients, much sooner.

Competitive markets, however, usually tolerate much less price discrimination than we should want. It is quite easy to package and market eflornithine in such a manner that a New York matron will not look to buy her face cream by mail order from a relief agency in Burkina Faso. But in more prosaic circumstances, discriminatory prices can be difficult to sustain. Typically, resellers step in to buy low and sell just a bit higher, until the price gap all but disappears. This problem alone often guarantees that drug companies can't maintain the pricing schemes that would allow them to recover the cost of developing the least-profitable segments of the potential market. So long as a drug remains under patent, its owner can limit that problem by tightly controlling the distribution of the drug down to the retail level. But even then, competitors can use much of the valuable information associated with the drug without copying the drug itself—hence the pioneer's curse.

That the innovation-propelling drug science and patient-side science are so tightly linked creates powerful synergies that accelerate the development of both. But these synergies also present unusual challenges when we try to frame appropriate intellectual property rights. The innovation in new lightbulbs, microprocessors, books, songs, and movies can be truly new, but all the know-how in our drugs is, in key respects, directly derived from the vast library of biochemical know-how created in a process that began long before humanity arrived on the scene and which continues to this day inside our bodies and the microbes that infect them. To spur innovation in this environment we need property rights crafted for a unique industry whose mission is to decode, mimic, and mirror what has already been invented by the greatest innovator of all: life itself.

———

CURRENT LAW ADDRESSES two different ways of hitching a free ride on drug-related know-how. Biochemists can clone the pioneer's drug. Lawyers can (in effect) photocopy the clinical data that got the original drug licensed, and resubmit it to license a generic clone. Two forms of intellectual property determine how soon each type of copying can begin.

A patent grants a monopoly for a fixed term, during which the drug's chemistry may not be cloned. In 1984, ostensibly to address the fact that drug companies were losing much of the value of their patents during the long process of moving drugs through the FDA, Washington codified an additional "data exclusivity" right. For a while, photocopy licensing is forbidden. Roughly speaking, this is a copyright for the drug's label, which is there to describe, directly or indirectly, the patient-side biochemical environment in which the drug will perform well.

The length of this copyright varies—it can last up to twelve years for monoclonal antibodies and other biologic drugs, five years for nonbiologics. Lower limits apply to data derived from clinical trials conducted in support of a sort-of-new use of a previously licensed drug; the sort-of may involve changes related to active ingredients, strength, dosage form, route of administration, or conditions of use. Slightly longer copyrights are granted for new pediatric trials.

The copyright gives drug companies an incentive to continue acquiring new clinical data to refine how a drug is used, or to secure new licenses for entirely new uses. But the 1984 law was in fact a step backward. Until then, only the manufacturers of antibiotics could get a clone licensed merely by demonstrating that it was "bioequivalent" to the already licensed original. The results of the clinical trials that secured the licenses for all other drugs were protected as trade secrets, which can last forever. By sidestepping trade secrets, the copyright lets competitors introduce generics much sooner.

———

THE EXISTING DATA exclusivity rights address the correct problem, but they fall far short of addressing it fully. As a result, the owners of the bodies that could supply the most valuable clinical data have almost no financial stake in the process, and may therefore be less willing to supply it. Pioneering drug companies and others who work out how the patient-side chemistry operates are rewarded too little, if at all. Early users of the drug pay too much. Companies and patients who arrive later free-ride too much.

We need, to begin with, a property right to promote the discovery and validation of biomarkers—objectively measurable biological characteristics on the surface of an HIV virus or a cancer cell, or in our livers, for example, that can be targeted or tracked for medically useful purposes. Isolating some molecular fragment of nature's code shouldn't suffice; the right should be linked to the development of a new biochemical operating manual that is good enough to contribute significantly to the development of new drugs, diagnostic tools, or treatment protocols.

Until recently, the main way to profit from the pursuit of new biomarker science was to incorporate it in your own new drug or diagnostic sniffer. But identifying a target molecule and working out what role it plays is enormously valuable in its own right. Sometimes the molecule—a flawed kinase enzyme, for example—provides the blueprint for the design of the drug or of the diagnostic sniffer. Or the molecule may instead be used directly in the manufacture of monoclonal antibodies. The discovery of disease-specific biomarkers can also significantly accelerate and thus lower the cost of licensing drugs developed thereafter to hit those targets. A better understanding of what an existing drug targets or what roles the target plays also helps guide the development of systematic, multidrug attacks on complex diseases. And such an understanding provides half of the information needed to introduce the old drug to a new disease. Gilla Kaplan played a key role in transforming thalidomide into good and profitable medicine centered on disorders propelled by cytokines. The repurposing of eflornithine to treat sleeping sickness was launched by Cyrus Bacchi, a scientist at Pace University, who discovered that the sleeping sickness parasite depends on the enzyme that the drug targets and persuaded the manufacturer to provide him with a sample to test against the parasite. As epidemiologist Robert Desowitz recounts, "Twenty-two years later, there is still a sense of excitement when Bacchi tells of how astonishingly effective [eflornithine] was in his experimental animals."

At least equally valuable are the discoveries of how specific genes, enzymes, and other biochemicals in the liver and elsewhere affect how drugs move through the body or are involved in certain types of side effects. These discoveries can make broad categories of drugs safer and more effective by providing systematic ways to distinguish patients who should use them from those who shouldn't. Drug manufacturers and countless patients owe a large debt, for example, to those who established that genetic variations in one group of enzymes cause some people to metabolize about thirty types of drugs too quickly, before the drugs have a chance to work, and cause others to metabolize them too slowly, so that drugs accumulate to toxic levels.

The market recognized the value of biomarker science long ago, and numerous attempts have been made to patent it. Thousands of individual genes have been patented, but how much (if any) competitive turf each of these patents may cover will be litigated for years. The patent law covers "methods" as well as chemical compositions and products, but the Supreme Court and the federal appellate court that handles patents have resisted granting patent protection to "abstract mental steps" or "laws of nature." In 2013, in a case involving Myriad Genetics' patent claims associated with two breast cancer genes, the Supreme Court ruled that "genes and the information they encode are not patent eligible . . . simply because they have been isolated from the surrounding genetic material." But novel diagnostic tests, therapies, and processes that incorporate such genes remain eligible for patents.

As the courts have recognized, any unduly broad property right that limits the use of basic biochemical science will foreclose too much innovation by others later on. The discovery of a promising target, for example, shouldn't be allowed to halt all competitive development of molecules that can detect or attack it. But when biomarker know-how links molecular effects to clinical ones, it has moved far beyond abstract cogitation or laws of nature. This kind of know-how, as we have seen, emerges from either the search for patterns in large, complex bodies of molecular and clinical data or from the in-depth study of biochemical processes in individual patients. And it is an essential complement to the know-how embedded in diagnostic sniffers and drugs.

The digital world presents similar issues, but there the intellectual property rights that are obviously needed are firmly in place. All software is a series of "abstract mental steps." So, too, is the logic etched onto the surface of a one-hundred-million-gate Intel microprocessor, and the performance of each individual gate on the chip is governed by the laws of semiconductor physics. Yet the software can be protected by both patents and copyright. And the three-dimensional maps of integrated circuits are protected by a form of copyright created in the United States in 1984. Biomarker know-how supplies the maps of the systems that will run the biochemical code embedded in diagnostic sniffers and drugs.

To get private money fully engaged in creating these maps, we need a new data exclusivity right to cover them. The right should be narrow—one for linking a key enzyme to sleeping sickness, another for linking it to facial hair—but it should also last a long time. The objective is to allow the market to spread the cost of developing biomarker know-how across the broad base of future beneficiaries.

Biomarker know-how will often be tested and validated by success in the clinical trials conducted by the pioneering developer of a new drug or sniffer, in which case the pioneer deserves both the data exclusivity right that the law already confers and this new biomarker property right. But current policies implicitly assume that all the biomarker science that matters will get developed this way, and that assumption is plainly false. Some parts of the science concern only human or microbial biology and can thus be developed and validated by researchers and not-for-profit entities that lack the resources to design, license, and market a drug or sniffer. Others are developed by companies that collect and analyze clinical responses to licensed drugs and by doctors who investigate off-label uses.

Current policies also force most of the cost of developing biomarker know-how to be shouldered by the patients who buy the pioneering drug while its patent lasts. The market will develop more pioneer drugs, and government paymasters will be more willing to accept them, if one important component of first-pill costs can be spread across many more patients. The early users already shoulder more risk; they shouldn't be paying so much more as well.

The agency that issues biomarker rights should also have backup authority to oversee a compulsory licensing process analogous to those that currently allow singers, for example, to perform songs composed by others. Licensing fees should be modest, and total licensing fees collected might be capped at a level commensurate with the cost of developing the biomarker-validating know-how and its role in creating new value thereafter. As airlines and digital markets demonstrate, markets are usually quite good at developing calibrated pricing schemes that accommodate the greatest possible number of customers at the lowest possible average price. But when private parties are unable to frame their own licensing agreements, a property right subject to compulsory licensing is probably the best we can do. There's a delicate economic balance to be struck here, in promoting a discrete type of creativity that can be very valuable to society, but only when it launches a broad range of follow-up innovation.

———

WHEN THE PIONEER'S data exclusivity rights expire, a generic drug currently gets licensed by showing that it works and is packaged and labeled in exactly the same way as the original. The copycat manufacturer can't even be sued for any failure to warn, the Supreme Court ruled in 2011, because it has no control over the label. Some courts have decided that the pioneer may be sued instead, even if it no longer sells the drug.

But the FDA-approved label is history, or soon will be. That scrap of paper is Washington's attempt to tell you and your doctor whether your body is bioequivalent to other bodies in which the drug has performed well in the past. Except in the simplest cases, paper can't convey even a tiny fraction of the information that we should be using to fit drugs to patients. The accurate fitting will emerge from pattern-matching search engines powered by constantly growing databases. They will contain far more information than is currently collected in FDA-scripted clinical trials.

At present, specific intellectual property rights protect only the clinical data that emerge from trials conducted under Washington-approved protocols for use in Washington's own licensing process. Both the 1984 data exclusivity right and a broader marketing exclusivity right included in the Orphan Drug Act are tied to the licensing of individual drugs. They make no attempt to secure private economic interest in the many other, much broader ways in which clinical data are now used to launch the design of new diagnostic sniffers and drugs and guide their use.

The 2011 NRC report on developing a new molecular taxonomy of diseases discussed earlier proposes a nonproprietary "information commons" and seems to suggest that Washington can expose all we need to know about America's biochemical code with government-funded pilot programs, mandates, the curtailment of at least some existing intellectual property rights (albeit while "clarif[ying]" "guidelines for intellectual property" and "address[ing] concerns" about their loss), "strong incentives for payers and providers," and quite possibly "require[d] participation [in exchange] for reimbursement of health care expenses." Data-sharing standards must be framed to discourage "proprietary databases for commercial intent." At the same time, however, the report strains to suggest that its proposals extend only to the sharing of "pre-competitive" (and implicitly, at least, also "post-competitive") information. The co-chair of the task force has independently suggested an alternative "I-give-and-I-get" deal for patients—you upload in exchange for the right to download.

Self-interest and altruism already impel many patients to share with others the biochemical knowledge gleaned from their bodies. But to explore the biochemical frontiers, clinicians, patients, and healthy citizens will have to measure and track far more information than they routinely would. Plucking meaningful patterns out of complex floods of data is expensive. The private money won't generate valuable information without some robust right to own and profit from it. And many Americans will recoil at the suggestion that their biochemical data will be uploaded into computers ultimately overseen by a government that has an uneven record in limiting access to its own diplomatic cables and military computers. We need instead policies that will give drug companies and many other private entities strong incentives to collect much more data from many more patients, analyze it using the full power of the most advanced digital technology, and convey the information, in suitably synoptic form, to doctors and patients.

We should start by reaffirming the patient's right not to disclose information about his or her innards to anyone—along with a right to give it away, share it, sell it outright, or license it selectively, at whatever price the market will bear. The individual patient won't often be interested in haggling over what a clinical record might be worth to a hospital, insurance company, or anyone else. But affirming the private ownership of private information is the first, essential step in creating robust secondary markets served by those who would help collect and analyze the data.

Drug companies have shared interests in analyzing clinical records to find biomarkers that help predict metabolic processes and toxic side effects that affect many drugs, and ought to be encouraged to pool their resources in tracking them down and allowed to profit by doing so. Life insurance companies would be interested in biomarkers that foreshadow grave health problems long before they materialize—and the smart insurers would then find ways to continue writing policies while helping develop the science that would keep policy holders alive longer than they expected. Any insurer that covers long-term disability care has an interest in the development of biomarker know-how that could hasten the development of Alzheimer's drugs. And standing by to compete directly or provide the raw computing power are a pack of companies that specialize in aggregating and mining massive amounts of complex data, and making money while they do.

Market forces, properly engaged, would recognize that the value of clinical data varies greatly across biochemical space and time. The clinical experience of certain patients with unusual pairs or clusters of genes can be exceptionally valuable. The early users of a new drug contribute more to the mapping of the disease and relevant bystander chemistry than do those who use the drug later. Users of a fundamentally new type of drug contribute more than later users of a me-too drug aimed at the same target. Databases are more valuable if they contain more data from a more diverse group of patients who have used a drug for a longer period of time. Clinical records generated in trials of a drug that was never licensed because it was found to cause nasty side effects can be mined for information about what not to target and for what they reveal about the biochemistry of innocent molecular bystanders.

Information is very hard to contain, but as iTunes and Kindle demonstrate, markets can be created to contain it enough to keep its creation and distribution reasonably profitable. And the market has a long history of finding ways to maximize the value of information by consolidating it efficiently and providing broad access. For well over a century, competing companies have been pooling their patents and interconnecting their telegraph, phone, and data networks, online reservation systems, and countless other valuable information conduits and repositories. Today's aggregators buy up portfolios of patents and licenses for songs, movies, TV shows, and electronic books, and then offer access, in many different, economically efficient ways, to all comers. The global financial network depends on instant data sharing among cooperating competitors. Most online commercial services today hinge on the market's willingness to interconnect and exchange information stored in secure, proprietary databases. It is Washington that has been the persistent (and often inept) laggard in moving the IRS, air traffic control, and almost all the rest of the federal government's supervisory, transactional, and information gathering and dissemination into the digital age. Washington started trying to push digital technology into the medical arena around 2005; the results so far don't look promising.* And as we have seen, President Obama's own expert advisory council concluded in 2012 that the FDA's information technologies are "woefully inadequate."

The economics of the market for clinical data will, in all likelihood, be largely self-regulating. The accumulation of new data points steadily dilutes the value of the old, though their aggregate value continues to grow. New data can be acquired every time a patient is treated and monitored. The more widely a drug is used, the faster data will accumulate. The best way to accelerate the process that ends up making information ubiquitously and cheaply available is to begin with economic incentives that reward those who develop new information faster and come up with new ways to distribute it widely at economically efficient prices.

The emergence of separate markets for molecular and clinical information will also introduce an important new element of competition to drug markets. Among other things, the new markets will encourage independent com-

---

* In 2005, a Rand Corporation analysis concluded that the adoption of digital health records would save the United States at least $81 billion a year, and improve care, too. Washington has since spent billions of dollars to help move hospitals and doctors into the digital age. A follow-up Rand report released in early 2013 concluded, in the words of one of its authors, that "we've not achieved the productivity and quality benefits that are unquestionably there for the taking." Britain's National Health Service launched its $20 billion Connecting for Health project in 2002; nine years later, a parliamentary committee concluded that the NHS (and by implication the U.S. company that the NHS had enlisted to do much of the work) lacked the capacity to implement its "unworkable" schemes, and the project was canceled.

panies to anticipate or follow up on much of what drug companies currently do to get new diagnostic sniffers and drugs licensed. Independent companies will compete to provide the molecular science needed to design successful clinical trials of new drugs, and then promote the ongoing search for information about how licensed drugs perform or might be used in new ways. And as discussed earlier, a continuous learning process sets the stage for a dynamic licensing process that tailors the permitted uses of a drug to the most current, reliable knowledge about its likely effects in prospective patients.

Drug companies will inevitably continue to play a large role in the search for patient-side know-how, if only because much of it will be drug-specific information developed (ideally) during an adaptive drug licensing process. But for much the same reason that software companies welcome advances in digital hardware, drug companies should anticipate and welcome the rise of independent companies with expertise in connecting molecular effects to clinical ones in human bodies. Independent companies are permitted to interact much more freely with hospitals, doctors, and patients. And by accelerating the development and wider distribution of information about the molecular roots of health and disease, they will help mobilize demand for the expeditious delivery of drugs, the molecular antidotes.

For their part, doctors and patients will of course require access to the databases and predictive engines that provide the guidance needed to prescribe the drugs well. Market forces should nevertheless be permitted to determine the price of the drug and the separate price of the information that guides its use. Various schemes can be developed to let this happen while ensuring that the drug is always sold in conjunction with appropriate guidance. The code embedded in the drug's chemistry doesn't change, but our understanding of how it interacts with human bodies should continuously improve—even after the patent on the drug has expired, generics have flooded the market, and the original manufacturer has perhaps abandoned it. The important thing is to maintain incentives for the ongoing study of how a drug is performing, or might be used, while spreading the costs across the full range of patients who benefit from the continuous improvement in our understanding of how a drug can be prescribed well.

Despite much urging to the contrary in the NRC report and elsewhere, the databases and engines themselves should, when developed by private money, remain proprietary and protected by trade secret laws. Here yet again, the case for mandated sharing and free-for-all access is easily made— free always sounds better, at least at first. But the mandatory and free solutions, if they get developed at all, have a long record of lagging far behind what private money and competitive markets can do.

Myriad Genetics, the company that has patented two breast cancer genes, provides a diagnostic testing service; as of late 2012 the company had tested nearly a million patients, had compiled genetic data from the tests, and at last report was earning $400 million a year. Many other companies are already vying to combine large, efficient DNA sequencing systems with interpretive engines that can analyze many more genes and the associated databases needed to understand their clinical implications, at prices that rival what Myriad is charging to analyze just two. The NRC's recent report, excellent though it is on the scientific substance, ends up with a policy proposal to have government take full control of the already innovative, vibrant, and competitive market that the private sector is racing to develop.

FINALLY, WE SHOULD also allow market forces to develop reasonable schemes that integrate the development of drug science with the sale of drugs for treatment.

On one hand, patients, or their health insurers, must be allowed to start paying for drugs prescribed during the course of adaptive trials. Manufacturing drugs in small quantities, particularly the large-molecule mAbs, can be very expensive, and the small biotech companies that do much of the pioneering work may not be able to afford the process on their own. Rules drafted in the HIV-driven 1980s (and still on the books) grudgingly allow manufacturers to charge patients for the cost of manufacturing drugs distributed under the treat-investigate licenses. We should get over the grudging. Pay-for-performance schemes, already used in Europe, should be allowed here, too. Reasonable procedures, overseen by neutral arbitrators applying objective criteria, can surely be developed to administer such agreements. People rich enough to pour money into experimental medical therapies can afford good advice, too. Medicine can learn as much by treating the rich as by treating the poor. The visceral antipathy to the idea that drug companies might profit by selling drugs of uncertain value needs to be balanced against the fact that the uncertainty reflects, above all, the inherent complexity of human biochemistry—and also against the fact that the uncertainty recedes as experience accumulates. Many patients and competing drug companies will often start benefiting from what is learned about a good drug well before Washington is fully satisfied that the science is a wrap.

Conversely, patients should have the right to set their own price for what drug companies must pay to study how their bodies respond to a drug, both before and after it's licensed. Current rules allow only nominal sums set by Washington—the fear is that money will tempt patients to take risks that they shouldn't. But that is why doctors and other advisors must remain part of the process. So long as they do, money can play a positive role here as well.

To begin with, patients with unusual biochemical profiles sometimes own an extraordinarily valuable asset—an asset that may well kill them but could help save countless others. In an isolated fishing village in Venezuela, members of an extended family suffer from a very high incidence of Huntington's disease, a genetic disorder that afflicts about one in ten thousand Americans and leads to uncontrollable jerky movements, personality changes, and psychiatric problems. In another genetically isolated community in the Andes region of Colombia, many members of a clan of about five thousand people have inherited a single genetic mutation that guarantees they will develop early-onset Alzheimer's. By participating in drug trials, families such as these may well play a major role in delivering billions of healthier years to others, and billions of dollars to drug companies. Offering such families some economic reward to encourage eager participation in the development of invaluable medical science seems only fair.

More generally, the advance of molecular medicine now depends on acquiring much more data from many more individuals, both the sick, whether or not they are currently being treated, and the healthy who aren't. Science can't home in on which molecular factors matter without also gathering lots of information about those that don't. "Two of the most daunting bottlenecks" in the advance of molecular medicine, says Dr. Desmond-Hellmann, "are the low rates of participation in clinical trials and the current emphasis on patient privacy in research and clinical settings." Markets know how to solve such problems. There is much reason to doubt that government incentives and mandates can or will. Indeed, it seems unlikely that a federal mandate to participate in an insurance system in which Washington minutely regulates the diagnostic tests administered and the treatments provided, and then mandates the sharing of data acquired from each patient's body, can withstand constitutional scrutiny.

———

POLICIES THAT FAIL to align intellectual property rights with how new biochemical know-how is developed and used, flattening prices or loading too much of the front-end costs on early users, suppress the many economically efficient opportunities to let the rich subsidize the poor. Such policies promote small, shortsighted reductions in what we spend on current consumption in ways that deter private investment in knowledge-based assets that last forever and will deliver far larger savings over the long term.

Class envy is even more economically toxic here than elsewhere. Of course desperate patients who clamor to be included in drug trials are hoping to be cured themselves, but often they are fully aware that the drug is more likely to fail than deliver a miraculous recovery, and they are motivated at least as much by a determination to transform some small part of their own suffering into salvation for others. And if they happen to have led lives so industrious or fortunate that they are now covered by Cadillac insurance policies and can afford Tiffany drugs, so much the better. In doing all they can to help themselves, they can give much to medical and economic posterity.

The crowd doctors make health care fair by making it flat; the molecule doctors exploit the peaks and valleys of Vaniqa economics. Crowd doctors make collective calls about safety, efficacy, cost, and insurance coverage; molecule doctors profit by pursuing diversity down to the biochemical bottom and beating it one molecule at a time. Flat drug prices are not good for us; price spreads wide enough to cover first-pill costs and serve last-pill pocketbooks are good for us.

Washington launched the modern era of comprehensive biochemical mapping when it started funding the Human Genome Project in the late 1980s. Rarely has Washington found a better way to spend $3 billion of taxpayer money. But far more mapping has been done since then with private money. Finishing the job is going to cost trillions, not billions—far more than Washington can afford to pay. We must give the market the property rights and pricing flexibility that will get the private money fully engaged and leading the way.

Part Four

THE CULTURE OF LIFE

17

DYING ALONE

"IN THE VICTORIAN version of the Puritan ethic," historian Gertrude Himmelfarb writes in *The Moral Imagination*, "cleanliness was, if not next to godliness, at least next to industriousness and temperance." First they rebuilt the sewers, but with the emergence of germ science in the 1860s and 1870s, the Victorians came to understand that while public sanitation would eradicate many germs, escaping the rest required a robust germ-hating culture. By the time most of our grandparents were born, protecting public health had become, in large part, a private duty—one that included submitting to unwanted vaccination. We have spent the last half century not just abandoning that duty but turning it inside out. To a considerable extent, private health is now viewed as a public responsibility, so much so that the one health-related right that's sharply curtailed today is your private right to collaborate closely with the biochemists and doctors who might be able to help save you from yourself.

With timing that could hardly have been more perfectly wrong, the shift from the old culture to the new began just as the old, quintessentially public epidemics that spread especially fast and indiscriminately among children were giving way to the almost private epidemics of sexually transmitted diseases and to the altogether private diseases spawned by our genes and lifestyles, often in old age. Big, efficient, technocratic government agencies don't do virtue, sin, and personal responsibility; they requisition, subsidize, proscribe, and mandate. They teach—implicitly but persistently—that disease is the government's responsibility, not yours. The health-related social norms of the past now seem as antiquated as the whalebone hoops that defended Victorian chastity. Infectious diseases provide the starkest proof of the cultural shift. Washington now controls a huge germ-defense complex that lacks the essential fuel of a social and political culture needed to operate it effectively.

In practical terms, the problem can be addressed in one of two diametrically different ways. As discussed in this chapter, we can rail against irresponsible personal behavior, try to impose responsibility from the top down, and surely fail, because most Americans (I count myself among them) believe so strongly in leaving lifestyle choices to individuals. Or, as discussed in previous chapters, and further in Chapter 18, we can accelerate and expand individual access to the diagnostic devices and drugs that give individuals the power to scrutinize and control their own bodies.

———

THE MOST SECURE health care right in America today is the carefree right to mess with your own biochemical space pretty much as you please, so long as you go it alone and steer well clear of drug companies and doctors. And mess we do. Most of today's epidemics reflect individual choices, freely made. In the heyday of public health, public money went to clean up public filth. Today we spend our private money buying bad health by the pack, the bottle, and the Happy Meal, seducing disease in the bar and searching for it online. Slick marketing undoubtedly plays a role, but at the end of the day, the diseases don't come to us, we go to them. They are spread not by sewers but by our restless pursuit of things we crave.

Our right to pursue and indulge in such things is firmly anchored in the Constitution, or so a series of Supreme Court rulings seems to have established. Congress and many state legislatures have followed the Supreme Court's lead.

In the 1965 *Griswold* decision, discussed in Chapter 6, the Court discerned a right to prescribe and use the birth control pill in five different constitutional provisions, none of which, as soon became clear, provides any guidance on where the right begins or ends. A solid majority of the justices had no trouble rejecting the argument that Connecticut, the state in which the case originated, could regulate the pill because its availability might promote sexually transmitted diseases. That problem, they agreed, could be handled in other ways, because the state remained free to prosecute "sexual promiscuity or misconduct," including "fornication," "adultery, homosexuality and the like"—that was "beyond doubt." But before long the Constitution was protecting all of the above, along with abortion.

Meanwhile, the new sexual rights were quietly turning into a new infectious culture. It has taken a great deal of readily avoidable suffering and death to establish that people do need sexual taboos—taboos at the very least robust enough to thwart microbes, if not with less sex, then with more latex.

By the time HIV surfaced in 1981, a tragically large number of young women had contracted chlamydia infections serious enough to leave them infertile. Herpes, gonorrhea, syphilis, and two types of hepatitis were also on the rise, along with a raft of other bacterial, viral, fungal, protozoan, and parasitic infections. Other weakly infectious diseases—meningitis, for example—embraced the epidemic of physical intimacy. And the infections sometimes spawned other diseases powered by our own chemistry. Chronic hepatitis C is responsible for most cases of liver cancer, and while the main transmission route is exposure to contaminated blood, certain types of promiscuous sex also play a role. Like HIV, the virus lurks unnoticed for years, and it now kills more Americans than HIV does. About four out of five cancers of the cervix and other sexually active organs and orifices are linked to viral infections.

While the legacy germs were disappearing from sight and HIV was spreading unnoticed across the country, other parts of Washington spent the 1960s and 1970s proving that most health problems still originated out there, where the authorities could control them. There were serious issues to address—most notable were high exposures to hazardous chemicals in the workplace. But government agencies were soon busy concocting theories and compiling statistics—many of them wrong or wildly exaggerated, we now know—to prove that noxious workplaces and environmental pollution were responsible for a large fraction of the birth defects, cancers, immune system disorders, and other diseases that were surfacing as the legacy germs receded.

Trial lawyers launched a new wave of tort litigation to cash in. New standards of strict liability often excluded evidence that private behavior—smoking, for example—might have been the primary cause of a plaintiff's disease. With memories of thalidomide still fresh, drugs became attractive targets—particularly drugs used by seemingly healthy people. Something must have caused the mother's stroke, the birth defect, or the infant's brain damage. The oral contraceptive, morning sickness pill, or vaccine had arrived on the scene at about the right time. To many people, a sequence of events that coincides with strong preconceptions about how things work can be overwhelmingly suggestive.

Today, the new liberties also help shelter the homeless woman who sleeps next to the sewer and the mainliner who shares needles in the abandoned row house. Some states, though not all, have made it a crime to expose others to HIV, the most lethal sexually transmitted disease, but prosecutions are rare, and by and large an infectious lifestyle, once a crime, is now a constitutional right. To judge by the appalling trends in the incidence of sexually transmitted diseases (110 million Americans infected, 20 million new cases a year), it is a right that people of every sexual orientation, in growing numbers, continue to exercise in ways that harm themselves and others.

A constitution that secures bedroom rights almost certainly protects the kitchen and den, too. If you can get your hands on it legally, there's little doubt you have a right, at least in the privacy of your own home, to eat, drink, or smoke it. If you're especially vulnerable to certain risks and ought to take pains to avoid them, there's a good chance that Washington or your state capital has affirmed your right to put yourself in harm's way. Laws that bar discrimination against the disabled strain to keep both genetic and lifestyle frailties, such as obesity, from playing any role in the workplace and other commercial and public spaces. Pregnancy discrimination is illegal—the employer must make the chemical factory safe enough for the unborn baby, too. The mentally ill have a broad right to be released into "community-based" care. All too often, that protects their right to live with vermin and die slowly on the sidewalk.

Certain of these rights emerged from the new and in some instances controversial readings of the Constitution, but they have persisted and been expanded by federal and state laws because they reflect what are now widely shared values and a collective commitment to equal treatment and participation in public life. But they have other, mostly unintended effects, too. As noted in Chapter 1, humanity has a long history of blaming its diseases on outsiders, and it is reasonable to fear that even today, discussions of how diseases are linked to genetic and lifestyle factors will resurrect or reinforce old forms of prejudice and bigotry. But some of those links are indubitably real, and to beat the diseases we must understand and be willing to confront the causes. Similarly, we have found many ways over the years to harm ourselves with workplace toxins and bad drugs. But here, too, blaming diseases on such causes indiscriminately, when many of the real causes in fact lie within, reinforces the view that someone else will have to deal with them. Or, worse still, it blurs the line between causes and cures to the point where many people stop using readily available defenses against serious hazards—as happened with vaccines.

Establishing private rights to behave in ways that undermine health while broadly reinforcing the view that the real threats to our health lie elsewhere and should be controlled by others has helped lead to a dreadfully wrongheaded change of course in our collective pursuit of better health. Our grandparents focused on prevention. Better sewers and universal vaccination didn't help those who were already infected but did protect their healthy neighbors, and would continue to protect the community for generations to come. Charities became actively involved in promoting the development of new vaccines. Inspired by Jonas Salk's public vaccination of his own children with the polio vaccine he had developed, other families eagerly enlisted their children in early field trials. Most people celebrated the arrival of new vaccines. Today we specialize in the reactionary medicine of last-ditch drugs administered to gravely ill patients. We do so because Washington, its endless talk about prevention notwithstanding, vigorously affirms our private right to live for today, and has made at-the-brink medicine much more attractive to drug companies than the look-ahead alternatives that molecular medicine could provide.

As GOOD DOCTORS will often tell you, the best antidote to a lifestyle disease is a lifestyle cure. Don't smoke. Eat right. Use a condom. Just say no. But the Constitution affirms the private right to just say yes, and doctors often confront people who insist they're perfectly healthy, in their own, personal way, and just want to be left alone. Quite a few make their choices with middle finger firmly flipped in the general direction of the universal nanny. Many smokers still strongly believe in their right to poison themselves. In growing numbers, the overweight don't want to be tagged as sick and hectored to eat less; they just want courtesy and respect.

Even attempts to tell people that they have a problem can arouse fierce opposition. When Washington officially added HIV to its list of sexually transmitted diseases, the development of effective treatments still lay a decade in the future. The only way to fight the virus was to prevent its spread, which meant preventing infectious behavior. For twenty-five years, however, Washington, under intense pressure from various advocacy groups, doggedly opposed policies centered on promoting or even permitting voluntary self-diagnosis with home-test HIV kits. By the time federal authorities finally reversed course, about one million Americans were HIV positive. The CDC estimates that two hundred thousand Americans are currently infected but don't know it. Washington now says that HIV screening should be included in routine checkups.

Lifestyle doctoring gets even more delicate when medicine discovers genetic factors entangled with the lifestyle diseases. Some genes predispose carriers to nicotine addiction. Others have been linked to about half of the risk of alcohol addiction, others to obesity. The overweight, among others, sometimes welcome the opportunity to blame their genes, rather than bad choices, for their condition. Some gay-rights advocates worry that the discovery of gay genes would legitimize talk of "cures." Prenatal testing now gives parents the power to select for children who suffer from deafness or dwarfism, because they want children who live and enjoy life as they do. What some characterize as "the deliberate crippling of children," others see as celebrating their own diversity as they exercise their constitutional right to control their own procreation.

Mindful of how delicate the issue of treatment can become, the lifestyle doctors sometimes try to go after the suppliers of lifestyles instead. But many people earn their living selling what doctors don't want us to buy. Washington continued to subsidize tobacco farmers for years after the surgeon general began exhorting us all not to smoke. While First Lady Michelle Obama was denouncing fattening food sold in restaurants, and one office in the Department of Agriculture was directing Washington's antiobesity campaign, another office down the hall was helping Domino's develop a new, super-cheesy line of pizzas and was also covering the $12 million cost of a new marketing campaign.

Whenever it decides to blame the pizza rather than the eater, Washington reinforces the view that the problem is out there, not in here. We don't require the private citizen to scrutinize his genes, family history, or cholesterol count before he enters the pizzeria—the Federal Office of Quiche and Pizza will make sure that there's less cholesterol in the pie. If FOQP fails, some other federal acronym will see to it that someone else pays to suppress the cholesterol in his blood or unclog his arteries. Regulators strain to prevent insurers from scrutinizing genes, cholesterol levels, or other risk factors before they write policies. The push now is to require insurers to cover even arteries that are terminally clogged before the patient applies for coverage.

Pie control isn't going to work. People who crave cholesterol will always find a way to eat more pizza or go with the cheesecake for dessert. Lifestyle doctors have been at it for a very long time but have made dismally little progress in engineering healthier lifestyles—and by many indications they continue to lose ground. Prohibition didn't prohibit much or last long. The unhealthy consequences of bad diets have been known for almost as long as moms have been moms, and public authorities have echoed the moms throughout the decades when obesity surged. Smoking rates among women began ramping up sharply soon after public health authorities began begging everyone not to smoke. The rates have been declining since, but change came very slowly. That a handful of companies dominate the tobacco market, and peddle their poison in a limited number of easily identified packages, simplified the assault on their wares. No other killer-lifestyle market presents such a simple, clear target.

The lifestyle doctors will fail because they are trying to cure the lifestyle diversity of a free, open society, populated by immigrants from around the world. Our freedom doesn't melt away our lifestyle differences; it unleashes and amplifies them. It doesn't lose genes or blend them into a uniform biochemical smoothie; it remixes them to create a far broader range of biochemical profiles than any found in insular, xenophobic societies. People can indeed learn to make smarter choices. Some do, but many others don't. Some don't wish to. Some just can't seem to stick with healthier habits, much as they wish they could, and try as they may.

For a large and probably growing fraction of America's population, beating gene-lifestyle diseases will therefore come down to a choice between wishful lifestyle doctoring that doesn't work and drugs that do. After years of recommending diet and exercise as the starting point in treating diabetes, most doctors now agree that patients should start on drugs at the first sign of high blood sugar, because the consequences of a couple of years of failed lifestyle therapy are too dire. "Whatever your thoughts are on the importance of self-control and willpower and profligacy, and that we shouldn't be such pigs, that we should exercise more," observes Dr. David M. Nathan, director of the diabetes center at Massachusetts General Hospital, "the truth is that we are what we are."

––––––––––

TODAY'S PHARMACY IS quite well stocked with diabetes drugs, along with a handful of look-ahead drugs that can intercept some of the other perils of bad diets early on, before they trigger a heart attack or stroke. But obesity and smoking propel a slew of processes that sicken and kill different people in many different ways. Obesity is linked to heart attacks, strokes, and a substantial fraction of the most common cancers, along with arthritis, liver and gallbladder diseases, and a host of other problems. The smoking-induced cancer may end up in the patient's lung, larynx, esophagus, stomach, pancreas, cervix, bladder, or bone marrow, and smoking is associated with so many other disorders that it can be viewed as an all-purpose accelerant of aging.

At present we respond mainly with expensive, last-ditch medicine often administered too late to help. To beat most of the problems rooted in unhealthy lifestyles we will probably have to intercept them much earlier. But drug companies are spending far less on the development of look-ahead medicine than the pressing need and evident demand would justify, and the preventive drugs are lagging far behind the late-stage, reactionary drugs. In 2004, the FDA issued a report on "the widening gap between scientific discoveries that have unlocked the potential to prevent and cure some of today's biggest killers, such as diabetes, cancer, and Alzheimer's, and their translation into innovative medical treatments." The gap is still widening.

Drugs that attempt to control unhealthy behavior rather than its consequences aren't doing any better. The unmet demand has created a huge market for appetite suppressants extracted from herbs and other natural sources, which are available because "dietary supplements" are regulated much less strictly than drugs. Many patients resort to fixes that are widely available though illegal. Some find their way to doctors who concoct appetite-suppressing cocktails using off-label prescriptions of drugs licensed to treat seizures, opiate overdoses, attention-deficit disorder, depression, diabetes, sleep disorders, and smoking.

We know why the drug industry, so often accused of pursuing nothing but profits, has been so slow to start cashing in on the huge opportunities presented by lifestyle diseases. The FDA, the tort system, and the vaccine market taught most drug companies everything they care to learn about the perils of look-ahead medicine. Drugs that are used by the seemingly healthy are hazardous to the health of private money. In Washington, the politically safe path is to say no to drugs for the suspiciously healthy and yes to drugs for the desperately ill, and that's what Washington does. If your genes or lifestyle are going to kill you someday, hurry up and wait. Washington won't hurry until it's clear that they're going to kill you quite soon.

So long as the regulatory focus remains rigidly tied to the clinical symptoms, look-ahead drugs will, under current FDA rules, have to be tested in enormous crowds and will take decades to evaluate. But most will perform poorly because one size won't fit all, not when drugs aim to control behavior or tiny biochemical seeds of biochemically complex problems buried deep inside our bodies. When they intervene early and are used for years, the drugs may end up being blamed for countless side effects, perhaps some that they do in fact cause, but also many that they don't. Any drug that directly affects behavior can be blamed for any number of different effects, good and bad, that Washington's regulators and paymasters can balance as they please. The tort system compounds the problem. Almost anything can happen when cause-and-effect questions are litigated again and again, years later, in front of juries scattered across the country.

Drugs that intervene in the late stages of grave diseases, by contrast, can be licensed quite quickly if they are even modestly effective. Juries will be less easily persuaded that the patient, already gravely ill, would have fared better without the drug. The drug company can set a high price and has a good chance of recovering its up-front investment quickly . . . at least until the government's paymasters pile on; then the stage is set for complete paralysis. They can't afford last-ditch medicine for all, and will have little trouble explaining why they shouldn't pay for the small gains that the introduction of each new last-ditch drug is likely to deliver on its own.

Thus, between the patient and the drug—whether it's a look-ahead drug or a late-stage one—stand various forces, all there to help the patient, none able to deliver the drug, but each with the power and a reason to see to it that no one else does, either.

———————

MISSING FROM THE picture is the patient who is willing to get out ahead of Washington, limit or forgo any future right to sue, and—cautiously, of course, and with careful monitoring—start using a look-ahead drug today in an attempt to head off some peril lurking in his or her quite distant medical future. The unwilling patient, on the other hand, is in fine shape—the right not to take even the Washington-approved drugs that would improve unhealthy biochemistry is as secure as the right to enjoy super-cheesy pizza.

Begin with the unwilling. That others will eventually share the higher cost of treating the stroke or heart attack that cheap drugs might readily have prevented seems vaguely wrong—but surely not wrong enough for us to summon the health police to force the daily Lipitor down our neighbor's fat throat.

Private right has eclipsed even a person's duty to protect the rest of us from his or her infectious body. Many adults just don't bother to get vaccinated against influenza, pneumonia, and hepatitis. Nineteen states now accept unvaccinated children in the public schools if their parents register "philosophical objections" to the shots. After a long period in which vaccination rates fell steadily, they recently started inching back up in most of the country, but remain dangerously low in areas of the Northwest, and significant numbers of parents now delay or reject one or more of the recommended childhood vaccines. Whooping cough and measles have recently been killing people at rates not seen since the 1950s. The first victims are often babies, who are still too young to have been fully immunized, or the unborn, who can develop serious birth defects when their mothers contract rubella. Also at risk are the elderly and others whose immune systems are too weak to handle vaccines. In the words of Paul Offit, chief of the division of infectious diseases at the Children's Hospital of Philadelphia, medicine now confronts what many view as an "inalienable right to catch and transmit potentially fatal infections."

Meanwhile, the germs have diligently continued to immunize themselves against our antibiotics. Stubbornly persistent pelvic infections, along with drug-resistant gonorrhea and syphilis, soon joined the other enthusiastic celebrants of the sexual revolution in the 1960s and 1970s. Other drug-resistant germs then emerged in the septic underclass. Destitute, needle-sharing, and often HIV-positive patients who didn't strictly follow the complex, unpleasant drug regimen used to suppress the virus became human petri dishes, in which other microbes multiplied and evolved to resist the stew of antibiotics that were then prescribed as a last resort.

Many of the rest of us have become part of this problem, too. Insured to the hilt, we arrive at the doctor's office demanding a cure for our nasty cold, and the doctor, who knows that there is none, prescribes the Levaquin anyway, because we won't be happy otherwise and someone else is paying for it. Antibiotic gluttony is a new lifestyle disease, an indulgence that almost all of us can enjoy, often at Washington's expense. There are few philosophical objectors here. Fevers, rashes, mucus, and pus, right now, in your very own body, help keep philosophy in perspective.

When the government still tries to clean up human bodies with the same heavy hand that it uses to clean up leaky sewers, it can end up fighting both those who have forgotten the epidemics of the past and those who remember them too well. The forgetful are sick and tired of being hectored by some distant nanny about washing hands, vaccinating kids, and countless other time-wasting nuisances. Those who are not forgetful seem to believe that the little routines and habits of daily life are too important to entrust to a distant nanny, least of all a nanny that preaches health but dares not criticize promiscuous sex.

Consider, for example, a vaccine licensed in 2006 to protect against the human papillomavirus (HPV), which is responsible for about twenty-six thousand cancers a year. Developed by scientists at the National Cancer Institute, it's designated a "childhood" vaccine, which helps shelter it from the tort lawyers. To be effective, however, vaccination must occur before exposure to the virus, and on average, each new sexual partner exposes a girl to a 15 percent chance of infection. The Centers for Disease Control therefore declares that all girls should be vaccinated before. That, the federal authorities have concluded, would be at the age of eleven or twelve. After some debate, Washington also began recommending vaccination of all boys starting at age nine, because of the rising incidence of HPV-related cancer of the esophagus and anus in young males.

Quite a few parents have concluded that the federal authorities can go to hell. The forgetful are beyond help; they're probably skipping other vaccines, too. Many other parents are probably holding fast to a faith and a culture that still seek to protect children from sex itself. People who believe government can achieve anything will say that it should just have handled the HPV vaccine more delicately. Perhaps—but the fact is, the germ police have ended up at loggerheads with the people they most need as their closest allies: parents who teach the taboos and rules that provide a crucial first line of defense against the planet's most persistent and clever killers of children.

Even when it doesn't reach the point of turning parents against vaccines and breeding new germs that we don't know how to kill, the government takeover has left many people with a triple sense of entitlement—to germ-free life, risk-free drugs, and wallet-free insurance. With the drugs all selected and minutely controlled, regulated, and paid for by others, most of us don't have to give a moment's thought to who developed and manufactured them, still less to whether the suppliers are hard at work developing new and better ones to protect us against germs that don't yet exist.

Over our morning coffee and toast, consumed in our tidy little kitchens, we read that drug-resistant tuberculosis is a cause for growing concern—but mainly in prisons. So are new drug-resistant staph infections—in tattoo parlors and the foulest of locker rooms. And it's in private drug dens, bedrooms, and bathhouses, of course, that infectious germs have made their biggest comeback, contriving to get themselves spread by not-quite-private needles and genitalia. True, the germs incubated in abandoned houses, cardboard boxes, and other hovels have drifted into run-down urban hospitals, whose emergency rooms often provide primary care to the patients most likely to harbor the worst germs. But they haven't moved much farther than that, so they aren't our problem. Not yet.

The germs couldn't be happier. In Washington, they cheer on the FDA. In tattoo parlors, locker rooms, drug dens, brothels, and other sepulchers of freedom, they celebrate our Constitution. In nurseries and nursing homes, they raise a raucous toast to our trial lawyers and philosophers.

THAT BRINGS US, finally, to the willing, the people who, like their grandparents, are still eager to look to the future and take control of the disease before it takes control of their bodies, improving their lifestyles if they can, and using drugs if they must. They may turn to drugs simply because they recognize that they, or their children, just aren't self-disciplined enough to live as healthily as they should but are wise enough to use a drug or vaccine prophylactically if one is available. Or they may know that they carry toxic genes that make drugs the only alternative. Either way, they refuse to feel like victims. They blame themselves. And then they set about finding a solution by mobilizing the political and social forces required to unleash the enormous power of modern molecular medicine, as the gay community did so successfully in the early years of the battle against HIV.

We owe a significant fraction of our current life spans to such people. They were the ones who accepted the risks of the early vaccines, antibiotics, and antiviral drugs, and by doing so, they eradicated germs, built confidence in the antidotes, and helped develop better ones. Others played a similar role by taking the drugs that arrived later to start taming human chemistry. They supplied the information that allowed us to start understanding the code that makes us what we are, molecular warts and all, but also makes us smart enough to look ahead and confront the enemy before it's too late.

They were able to fill this role because the law allowed them to. It wasn't as libertarian as Justice Peckham's law, but it did leave the market with enough room to assign risks and responsibilities clearly, sensibly, and definitively. By allowing both Washington itself and drug companies to make more cautious, limited promises to patients, and allowing well-informed patients to accept more uncertainty and risk, the law maintained an environment in which the private money was eager to develop look-ahead drugs.

In health matters, as in the rest of life, some people are more willing than others to think ahead and accept both the cost and the risk of acting today to make things better tomorrow. Because so much of the value of every drug is pure information, these people can shoulder the risks for countless others who are more cautious, improvident, or not yet born. They will also often be the ones with steady jobs, Cadillac insurance policies, or money in the bank to pay for expensive care out of their own pockets, however bankrupt Washington may be.

Much of the time, the people who are willing to get started on the still-uncertain medicine are also the only ones who know or could readily find out why they should get on with it sooner rather than later. The patient knows better than anyone else how likely he personally is to stick with a diet or be exposed to HIV, or how many of his close relatives died early of strokes or colon cancer. She alone can find out whether she carries a cluster of genes that make breast cancer very likely or Huntington's disease inevitable, or whether she carries the gene that hinders her brain's ability to prevent the accumulation of a protein that causes Alzheimer's.

For all our freedoms, the one freedom you don't have is the freedom to take full, actively engaged responsibility for your health. That freedom ends the moment you try to team up with the people most able and eager to develop and sell the diagnostic tools and drugs that might help you take control of infectious perils or your own chemistry before it's too late. Others will decide if, when, and on what terms you may accept the uncertain risks and pursue the possible benefits of what may be the antidote to the slow, tiny, and invisible peril that is going to kill you.

18

THE RIGHT TO SNIFF

On the long list of your constitutional and statutory rights to control your own body, one is especially conspicuous by its absence—your right to take a close, careful look deep down inside to find out what exactly you might wish to control. It's time to add that right to the list. Doing so, as discussed in previous chapters, will democratize access to the information on which all the rest of molecular medicine hinges. That could well have a profound impact on both individual behavior and the political and social factors that shape our health care policies. And it will greatly accelerate the advance of molecular medicine itself.

We will never catch up with nature, but we do have a pretty good of idea what we need to transform molecular medicine into a science. We need molecular profiles of the microbes that make us sick, as well as of the much larger number that dwell more or less peacefully inside us and help keep us healthy. And we need torrents of molecular data extracted from countless individual bodies—Stanford's two-year iPOPing of Professor Michael Snyder (Chapter 2) foreshadows what lies ahead.

And then still more torrents to track inanimate environmental factors that interact with life's biochemistry. One MIT researcher, for example, recently set up a network that uses sensors and software linked to smart-phones and calling records to track members of sixty families living on campus—where they went, whom they met, what they bought and ate, how much they weighed, how they felt, and many other aspects of behavior and health. The data revealed all sorts of things. By observing behavioral changes, to pick just one example, trackers can apparently discover that you have the flu before you know it yourself.

All such analyses will soon be linked to still other sources of data—reports of the weather, for example, which affects behavior and thus whatever behavior affects. Environmental sensors, like medical sniffers, are getting cheap, fast. They're already widely deployed to monitor and control the insides of homes, offices, factories, cars, ships, and planes, and to sniff the air for chemicals and microbes that might be released by terrorists. We are advancing rapidly toward a future in which we will be able to monitor continuously almost every factor that might affect our molecular health. And the massive computing power of the cloud will extract the patterns that can tell an individual how controlling—or failing to control—one or more aspects of molecular health today will affect his or her clinical health tomorrow.

The FDA has the authority to regulate it all, end to end, pursuant to its statutory duty to license any "instrument, apparatus, implement, machine, contrivance, implant, in vitro reagent, or other similar or related article, including any component, part, or accessory, which is . . . intended for use in the diagnosis of disease or other conditions, or in the cure, mitigation, treatment, or prevention of disease, in man or other animals." The agency can't possibly keep pace with the technology, still less with the torrents of information that it could deliver. As a licensing agency, however, it has the power to slow both down to Washington-approved speeds. But does it have the constitutional authority to do so?

————

TODAY'S WASHINGTON, WE learned earlier, enlists Bayes to persuade itself that women need fewer mammograms. Then Washington persuades itself that the rest of us aren't smart enough to deal with certain types of biochemical data, even if the data are perfectly accurate, and however carefully we are warned that their clinical implications are uncertain. There will be no iPOP in your bathroom until Washington is confident that the device will guide you toward the Washington-approved clinical truth.

Doctors have permission to sniff more than patients may, though they, too, are apparently so techno-credulous that Washington assumes they will ignore what they might otherwise have considered and charge ahead with medication or surgery if some misguided sniffer suggests they should. Many manufacturers are now racing to design implants that would do nothing but gather data that doctors would use to adjust drug dosages or antici-pate strokes, heart attacks, and other problems. But most manufacturers shy away from sharing the data directly with patients, because getting FDA approval to do so would be difficult and expensive. Medtronic's implantable heart defibrillator uses a wireless link to upload data from the patient's body to the company, which reports back to the doctor but not to the patient. The company is investigating the possibility of selling what it learns from these data streams—"the currency of the future," it calls them—to other providers and insurance companies, to help improve their ability to predict problems.

In its role as national censor of biochemical speech, the FDA usually sounds reasonable and acts sensibly. It regulates the kits that patients use to collect specimens—clumsy users, after all, might contaminate the samples they collect. It verifies that a sniffer actually performs as promised on the package—this many false positives, that many false negatives—when used as directed. It has licensed many reasonably reliable sniffers for over-the-counter sale, and quite a few more for use by doctors and labs.

But it doesn't stop there. The FDA agonizes over the possibility that patients who misunderstand what a sniffer reports may fail to get lifesaving treatment, or neglect to take proper steps to halt the spread of contagious disease, or overreact when they should leave well enough alone, or fail to start addressing right now problems that haven't caused trouble yet but probably will if the forewarned patients don't quickly mend their ways.

These behavioral judgments can't be solidly anchored in science, because any good or harm that a diagnostic test may cause depends on how patients or doctors will respond to one small piece of information in a constantly changing deluge of information. And any one-size-fits-all prediction on such subjects is bound to be wrong much of the time, because it depends on everything else the individual patient or doctor already knows, or might yet find out, about the sniff in question.

The FDA is often backed by members of the mainstream medical media—doctors, medical societies, insurance companies, consumer advocates, and representatives of the many government agencies that help pay for medical care—who don't hesitate to speculate about what might go wrong. One commonly voiced concern is that medical diagnostic kits are just a waste of money. "There is no need for one," declared *Better Homes and Gardens* in a 1966 discussion of home pregnancy kits. When the first such kit came to market a decade later, *Consumer Reports* dismissed it as "a needless purchase" for its readers. It also insinuated that the only people who needed it were promiscuous women with something to hide. By the 1980s, the vehement opposition to HIV test kits suggested that the main concern had shifted from too much privacy to too little. The FDA didn't even begin seriously considering the possibility of approving home HIV test kits until 1990, when (as one FDA official put it) the development of HIV-suppressing drugs "made it more important to identify people infected with the AIDS virus who might be reluctant to be tested at clinics or doctors' offices." The FDA could, of course, have said exactly the opposite—making diagnosis cheap and easy was most important when preventing further transmission was the only way to thwart the virus. Some years later, the FDA faced vocal protests for its refusal to license kits that would let parents test their kids for the use of illegal drugs. At congressional hearings, the agency found itself defending an internal memo's suggestion that such kits might foster family discord.

The FDA will always be able to cite fuzzy facts to support whatever it decides. There will always be plausible reasons to believe, and plausible experts on hand to attest, that information foolishly sniffed can be as perilous as crack. Wiser heads must decide what the rest of us may sniff because, well, they're wiser.

But however wise the heads, they aren't wise enough to tell us what all the sniffer reports might mean until many of the less wise hand over torrents of data for Washington to analyze. Even if it collects every possible data point acquired by every doctor whose bills it pays—a policy that would outrage many Americans and would quite possibly be judged unconstitutional—Washington will lack access to most of the relevant genetic and lifestyle data. Washington can't monitor and track the people who keep themselves too healthy to need much care, or wealthy enough to pay their own bills. Nor can it track diets, exercise, where people hang out, and all the ways they exchange molecular and cellular snippets of health and disease. And as we have seen, individual doctors, patients, and private companies that aggregate clinical data now lead the way in unraveling the molecular taxonomy of disease and have moved far out ahead of the FDA in assembling the databases and designing the analytical engines that fit drugs precisely to the patient-side chemistry.

———————

THE POLICE HAVE much broader rights to read your biochemical code than you do. Unless you're wearing a space suit, you leave a trail of microscopic biochemical debris wherever you go. Under the established finders-keepers rule for abandoned trash, fingerprints, blood, and the like, courts have consistently upheld the government's right to bioscan whatever it may find there. Suspects may find in their mailbox a postage-paid, lick-to-seal envelope with a claim form for a prize offered, they later learn, by the local crime lab.

The lab's bioscan may well reveal gender, race, and family connections. A close match with an innocent relative's DNA that happens to be on file can provide an excellent lead to the rapist, and genetic fingerprints correlate quite well with surnames. Ethnic origin markers in DNA scans are already being used to reconstruct facial features, and a quite accurate portrait can probably be conjured out of about five hundred facial and five hundred ancestry markers.

The Fifth Amendment won't be a problem. The right against self-incrimination, courts have long held, protects only confessions emanating from the conscious brain, not bullets, blood, or other incriminating evidence extracted from elsewhere in the body. The FDA won't be a problem, either. The FBI didn't need a licensed Walgreens test kit to sequence the DNA of the bacteria involved in the September 2001 anthrax attacks, or 23andMe to find a DNA match in anthrax spores obtained from an army research lab at Fort Detrick, fifty miles from Washington.

When the FDA tries to protect the rest of us from the perils of scrutinizing our own biochemistry, however, it should have to answer to the First Amendment. The Supreme Court has repeatedly concluded that "freedom of speech" includes a private right to listen, read, and study that is even broader than the right to speak, write, and teach. Enabling tools and technology—the printing press and ink that produce the newspaper, for example—enjoy the same constitutional protection. And as we have seen, courts have recently been affirming First Amendment rights to discuss off-label drug prescriptions. A constitutional right to make unapproved predictions about what a drug might do to a patient's body must surely also cover predictions about what the patient's own genes and proteins might do.

Washington will insist that sniffer licensing advances a compelling, life-and-death objective—compelling enough to meet the very strict constitutional standard that must be met to justify government regulation of speech. But Washington couldn't license television shows only for the edification of doctors, on the ground that *Grey's Anatomy* might impel some people to try silly things with scalpels. Nor could Washington limit a patient's right to read his own written medical records, nor even insist that he read them only with a doctor standing by to explain. The conventional speech-and-press landscape is littered with inaccurate, incomplete, and murderously wrongheaded medical advice, almost all of it constitutionally protected—though no one can doubt that such information sometimes leads to horrendously bad choices, such as rejecting the vaccine that would have saved a child's life. Biochemical texts are different, apparently, because we don't yet view them as texts at all, health is complicated, medicine is mystical, and the facts of life are too subtle and important to be sniffed with impunity by the masses. Under well-established constitutional principles, dodging the First Amendment requires more than that.

Today's sniffer licensing suffers from a further defect—it is, in the standard legal jargon, a blunderbuss "prior restraint" that discriminates among would-be users on the basis of broad-brush judgments about how wise they are and how they will react. Here, Washington's one-size-fits-all predictions approach pure speculation. Fallible sniffers in careless hands may indeed scare hypochondriacs half to death, but they may also save lives by supplying the sharp kick in the rear needed to propel the morbidly sanguine into a doctor's office. Even if a sniffer is always right, how patients or doctors will use it also depends on everything else that shapes their thoughts about disease, medicine, and life, including nonmedical interests in cost, convenience, privacy, and autonomy.

Neither science nor economics can objectively weigh and balance these factors even in the short term, and they can't possibly predict how that balance will shift as we democratize floods of new biochemical data over the longer term. When the FDA concludes that a sniffer doesn't really sniff the pregnancy protein or the HIV virus, the conclusion can be grounded in hard science. Sniffer licensing based on wiser-head analysis of how a woman is likely to react to the discovery that she is pregnant or HIV positive can't be. What that kind of licensing is most likely to do is promote a lazy, foolish reliance on others. Trust the sniffer if and when Washington says you can, and don't worry until Washington gets around to deciding how you should react to what it reports.

A CONSTITUTIONAL RIGHT to sniff can be affirmed without questioning the FDA's power to license drugs, which the courts have upheld against all "right-to-control-my-own-body" challenges. Drugs directly reshape the material world, in ways that can cause direct physical harm; sniffers only talk. Much of our First Amendment jurisprudence is centered, as it should be, on maintaining the line between the suppression of speech and the regulation of harmful conduct.

But not so fast. We now know that health, diseases, and their antidotes, which we viewed as opaquely physical when we didn't know any better, run on legible, logical scripts. And drugs now let us edit those scripts with deliberate, conscious precision. Gene therapies aim to change genetic conditions that haven't yet morphed into any disease, and before long they will be targeted at reproductive cells to shape a child's health before it's even conceived. Our grandchildren "will likely engineer themselves into what we would consider a new species, one with extraordinary capabilities, a *Homo evolutis*," predicts Juan Enriquez, a Harvard authority on life sciences and author of *As the Future Catches You*. We have in hand the tools to go there, if we choose to.

In the here and now, brain chemistry drugs take direct aim at molecules that shape memories, impulses, feelings, and thoughts. Uppers, downers, antipsychotics, memory enhancers and erasers, dementia drugs, opiates, and hallucinogens blur the boundary between body and mind. Students use Ritalin and Provigil to boost academic performance, and mainstream scientists and ethicists are now heard to argue that society should accommodate mushrooming demand for "cognitive enhancement." Some brain chemistry drugs are reviled by groups such as the Scientologists, whose members aspire to influence thoughts and feelings by more traditional (and constitutionally protected) means. Other drugs have the power to erase horrifying memories and return at least one corner of the psyche to the eternal sunshine of the spotless mind. Redirected against fond memories, they also show promise in the treatment of addictions.

Such drugs already have much in common with the patches we download to update the read-only memory of our computers—they are part conduct, but mainly they are expensive code that ends up stored on a cheap, almost irrelevant substrate. A newspaper subtly alters human chemistry from the top down, through the eyes, into the brain, and then via nerves and hormones to the rest of the body. A drug alters it from the bottom up, from the serotonin receptor up into higher tiers of activity in the brain. Viewed through biochemical spectacles, amphetamines are as good as a horror flick, and Viagra is just another dirty magazine. Down at the very bottom of life, the dualities of speech/conduct and mind/body disappear—texts printed in biochemicals control the bodies, and we can edit those brilliant texts pretty much as we please.

Meanwhile, science is moving us rapidly toward molecular medicine that must leave most of the control in the hands of doctors and patients, because they alone will have access to all the information about the patient's chemistry that is needed to prescribe drugs well. The debate about their right to take control no longer hinges on privacy, personal autonomy, or some other abstract legal right; at stake now is the delivery of the best possible medical care to the individual patient. The *Griswold* contraception ruling and the cases that followed don't begin to explain why constitutional penumbras should shelter one tiny corner of the biochemical infosphere but not the rest.

It's equally silly, however, to suppose that pragmatic judges will soon affirm an unbounded constitutional right to sell and buy the tools that will edit DNA and manipulate other parts of our private biochemical code as freely as Hollywood digitally cuts and splices movies. No Supreme Court opinion will soon articulate the grand constitutional theory that can deal with all the mind- (and body-) boggling possibilities that such a right would entail. We must unequivocally affirm the right to read biochemical code because the power to rewrite that code will inevitably be exercised, with unpredictable consequences that will dwarf every other revolution in information technology yet seen or imagined. If free societies are to maintain democratic control of their biochemical destinies, their citizens must be free to sniff all about the existential revolution as it unfolds.

19

THE END OF SOCIALIZED MEDICINE

A DOCTOR GOING door-to-door on New York's Lower East Side in 1911 could have cured rotting gums, grotesque swelling in the legs, brittle bones, noxious skin ailments, and a thick catalog of other debilitating diseases by distributing a pill—just one—that's now cheaper than candy, and so familiar we no longer view it as medicine. But a century ago, these terrible diseases killed and crippled so many people—tens of thousands of Americans every year—that many doctors viewed them as contagious. They weren't: industrial processing was stripping nutrients from the food that fed the city. Kids needed Flintstones.

The first vitamin was isolated in 1912. Others followed, and chemists soon found ways to extract or synthesize them cheaply. Prodded by private charities, medical associations, state health authorities, and federal guidelines, major food suppliers eradicated rickets, scurvy, goiter, beriberi, and pellagra by returning to their products what they had inadvertently removed; they also improved infant health enormously by fortifying flour, milk, and salt, and promoting the consumption of cod liver oil by pregnant women. By 1950, the Flintstoning of the American diet was routine, and the national menu was back to healthful again, or so many people thought.

They, too, were wrong. On June 19, 1987, Ben & Jerry's introduced Cherry Garcia, in honor of the man who played lead guitar for the Grateful Dead. The FDA struck back three months later when it approved the first of a new family of statin drugs that curb cholesterol production in the human liver. A synthetic statin licensed a decade later would become the most lucrative drug in history. At its peak, global sales of Lipitor were streaming $14 billion a year into Pfizer's coffers.

Let's not blame the victim. We don't choose Cherry Garcia; it chooses us. Lipitor is a lifesaver for six hundred thousand genetically unlucky Americans who harbor a bad-cholesterol gene or two on chromosome 19,

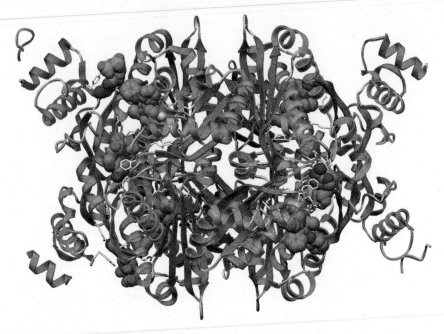

FIGURE 19.1   Computer model of Lipitor binding with a cholesterol-producing enzyme.
Source: http://www.sciencephoto.com/media/390143/enlarge.

and for another one hundred million helpless victims of our irresistible
supersize-me culture. Fourteen billion dollars is a bargain for problems as
pernicious as these.

Or is it? Let's blame the victim. The human body is so comfortable with
fat that it rarely complains about a cholesterol glut in the blood until seconds before things crash. Many who should be worried never even get their
blood checked. Many who do check it fail to take their Lipitor. None of us
really needs the pill anyway—just lose the ice cream, shed the pounds, stop
smoking, and exercise regularly.

Health once depended mainly on killing germs. Now most drugs are used
to tune people. Lipitor tunes our cholesterol. Anti-stroke medicines tune our
platelets, antidepressants our serotonin and dopamine, heart medicines our
angiotensin-converting enzymes, contraceptives our estrogen. Cancer drugs
tame or kill cells that harbor our own mutant genes. We have insulin for the
underperforming pancreas, clotting factors for bleedy blood, and Synthroid
for the tired thyroid. There were only a few dozen vaccines and antibiotics
in the field when the authorities declared the war on germs pretty much
won. There will be tens of thousands of people-tuners in the arsenal before

anyone is rash enough to announce that we have human chemistry fully under control. And the announcement will still come too soon.

Much of our need for those drugs stems from our own bad habits. By 2008, seven of the world's ten most profitable drugs owed most of their success to our foolish mouths. Two of those drugs lowered cholesterol, one suppressed the blood's tendency to clot on cholesterol plaques, one lowered high blood pressure caused in part by clogged arteries, two were for heartburn and acid reflux, and one was for asthma, which is often aggravated by cigarettes. In most Western countries, smoking still causes more deaths than all other readily preventable causes combined, but gluttony is catching up fast. Molecular medicine can also deliver thinner thighs, broader shoulders, fewer wrinkles, choreographed pregnancies, designer babies, better hair, more sex, restful sleep, and whatever else the incorrigible mountain biker may crave—and demand for such drugs is rising rapidly, too.

The demand, whether desperate or frivolous, will depend mainly on the details of our diverse, fragmented, personal biochemistry. And molecule by molecule, medicine is now making human code and its effects visible, predictable, and tractable. Cholesterol can be quite as lethal as cholera, but how much of it you have in your own blood this morning and what it will do to your heart in 2025 isn't a matter of dumb luck—not anymore.

Our rapidly improving ability to anticipate health problems and then control them makes health insurance—whether public or private—much trickier than it used to be. Recent political developments notwithstanding, the era of big government is over in medicine, too. Within a decade or two, if not sooner, a charismatic president will deliver on the promise to end health care as we know it.

––––––––

THE FIRST PATIENT we meet in Sherwin Nuland's *How We Die* is a powerfully built construction industry executive whose business success had "seduced him into patterns of living that we now know are suicidal," back when "smoking, red meat, and great slabs of bacon, butter, and belly were thought to be the risk-free rewards of achievement." The patient has been admitted to the hospital complaining of chest pain, but he seems fine and is resting comfortably. He's just seconds away from the violent heart attack that will kill him.

Cholesterol was one of the first molecules to emerge from the disassembly of lifestyle and gene statistics because it plays such a big role in gluing us together. In 1985, in their Nobel lecture on the lipid, Michael Brown and Joseph Goldstein noted that cholesterol was "the most highly decorated small molecule in biology," with eight prior Nobels on its shelf. But however often it may party in Stockholm, cholesterol will never be as simple as cholera, because what it does depends so much on whom it's gluing. Thin families are all alike, but every fat family is unhappy in its own way.

About one in three Americans officially has fat blood—cholesterol tagged somewhere between a trifle chubby and obese. The main cause is diet, but some bodies handle their Cherry Garcia worse than others. About two people in every thousand are born with a high-cholesterol gene. Their livers can't handle cholesterol as well as their mouths can, and they're twenty times more likely to suffer a heart attack before the age of sixty. About one in a million is born with two copies of the gene, and the heart attacks then begin in childhood. About one person in two of European ancestry is born with a gene for a bad heart, which boosts the risk of heart disease by 15 to 20 percent. About one in five is born with two copies, which more than doubles the risk. As Nuland notes, "The best assurance of longevity is to choose the right mother and father."

The second best is to find out where your personal gluts and genes will probably take you years from now, and order your life around that knowledge as well as possible. Molecular medicine's most important revelation suggests a very modest prescription: some things are best kept out of your lungs, guts, blood, and chromosomes. This, of course, is why we have rules to keep mold, dead insects, smoke, asbestos, benzene, radiation, and more out of food, workplaces, water, and air. Many individuals are now at least equally careful about what they eat, drink, and inhale, and some go so far as to tinker with genes. If each parent harbors a gene for high cholesterol, a couple faces a one-in-four risk of bearing a child who will have a disastrous cholesterol problem. A routine test early in pregnancy lets the parents find out. Most people, however, still prefer to take their genetic chances with offspring, and many still line up for a daily fix of fat, nicotine, tar, alcohol, salt, or worse.

That so many tests and pills can now help keep our chemistry in balance amplifies these differences. Health-careless people tend to be as casual with pills as they are with dessert. Lipitor only widens the gap between people who generally live informed, disciplined lives and those who don't.

———

IN THE "ONE America" vision of things, better government would deliver better diets and also more Lipitor to all, and that would make health care, perhaps even health itself, equitable and uniform. But however clear a health problem may be, and however simple and cheap the cure, molecular medicine is riddled with lines that the nanny state just can't cross. Who would ever have thought, for example, that the pursuit of thinner thighs might cause a thousand birth defects a year, many as dreadful as thalidomide's—and do it in the teeth of a federal scheme to save children from those very afflictions? In 1996, the FDA resolved to slip a certain drug into everyone's food. Dr. David Kessler, the FDA's head at the time, preferred to call it a "pharmacologically active ingredient," but what's the difference? Folic acid (vitamin B_9) promotes normal embryonic growth quite as powerfully as thalidomide subverts it—a healthy dose keeps a baby's spinal cord inside the spine, saving the child from a lifetime in a wheelchair. But the unborn baby needs a full ration in the first six weeks after conception, and roughly a quarter of young women have diets that fall short. So in 1998 Dr. K. directed that a modest dose be added to almost all flour, cornmeal, pasta, and rice touted as "enriched."

Doctors of jurisprudence, however, have concluded that no one can force an expectant mother to eat her risotto, least of all for the benefit of what may be growing in her womb. And on the advice of low-carb guru Dr. Atkins, millions of fertile young women stopped eating at Kessler's diner shortly after he adopted his stealth-health menu. The incidence of neural tube birth defects, which had been dropping rapidly before 1998, leveled off just as the FDA got involved. The public cure for a nutritional deficiency reached its limit, apparently, in a private cure for gluttony.

The spina bifida baby isn't responsible for his fate, but his mother certainly is in charge of hers—the Supreme Court says so. And many people, it seems, are reasonably comfortable inside their own skin, don't wish to change or medicate their lifestyles, and recoil at the thought of trying to jigger their children's genes. Stealth-health medicine worked brilliantly in the war against the legacy germs—by stemming their spread, vaccines and antibiotics protect the unvaccinated, too—but it can hardly touch gluts and genes. Freedom includes the freedom to burn your candle at both ends, even if it will not last the night.

———

So WHAT WILL insurers do with the pill that leaves the kick in a pack of Marlboros but magically neutralizes the poison? Will Aetna and the surgeon general both celebrate this miracle drug, congratulate Pfizer for racking up $40 billion in new sales in just one year, gracefully accept their respective shares of the bill, and watch calmly as smoking rates ramp back up? Will Congress declare that every smoker needs this drug, every smoker must get it, and Pfizer's price gouging must end at once? Or will some heartless bookkeeper in Hartford or Washington dare to suggest that enough is enough, smoking is foolish, and smokers can jolly well pay for the pill themselves—or, failing that, for their own cancer, emphysema, and heart disease? The Affordable Care Act takes one small step in that direction, in a provision that allows insurers to charge smokers (though not the obese) substantially more.

Lipitor for the Marlboro Man will take a while, but molecular medicine already raises questions like these every day, and they will keep piling up until they can no longer be concealed in the fine print of insurance policies or federal regulations. Common as they still are, and notwithstanding Washington's recently codified determination to give us more of the same, insurance systems that pool health risks indiscriminately are vestiges of the past. They can't survive what lies ahead.

Insurance makes sense for risks that individuals can neither anticipate nor control. It looks altogether different when your neighbor's problem is a persistent failure to take care of himself when he readily could if he chose to. Many people willing to share the burden of bad luck eventually tire of sharing the cost of what they view as bad behavior.

The new medicine certainly hasn't brought everything within our control—molecules don't predict car accidents and can't yet cure Huntington's disease, most cancers, or thousands of other rare, lethal disorders. A widely shared sense of common decency also impels protection of children and the elderly. In between, however, the unifying interest in health insurance is surely the sense that anyone can be struck out of the blue by a ruinously expensive health catastrophe. And step by relentless step, molecular medicine is taking luck out of the picture.

Now consider what that does to insurance economics. Most critics of the status quo focus on what they see as a problem that all health insurers face: runaway cost. But the real problem is that for many people, health care is getting *cheaper*. This is what makes actuaries wake up screaming in the night: disease is coming out of the molecular closet, and the new medicine splits health care economics in two. For the health-conscious, skipping the Cherry Garcia may be difficult, but it's cheap, and generic statins are cheaper than a heart attack. The health-careless skip only the pill, not the ice cream, and end up in desperate need of what helps the least and costs the most.

No one-size-fits-all, one-price insurance scheme can keep people happy forever on both sides of this ever-widening divide. Aetna can't offer uniform coverage to individuals who face radically different risks and who know it. Governments can't, either.

FOR NOW, PUBLIC authorities and private insurers conceal the growing cracks with Silly Putty definitions of what they'll pay for. The old medicine of scalpels and human services is quietly but firmly rationed, if not by ability to pay, then by making patients wait in ever-lengthening lines for access to overworked doctors, obsolescent labs, and deteriorating hospitals. The new medicine winds up rationed by slow-rolling many new drugs.

Britain, with one of the world's worst cholesterol problems, began prescribing statins years late, and much less aggressively than it should have. Nine years after the first statin had been licensed in the United States, Britain's National Health Service was still grappling with the fact that it couldn't afford the drugs, and its doctors were prescribing statins to only a small fraction of the people needing them. U.S. insurers, both public and private, often do much the same thing. They extend coverage to new medicines well after they're licensed, limit aggressive prescription early on, cover drugs sold in pharmacies much less generously than those administered in hospitals, and jigger deductibles and co-payments. The gap between what's covered and what the new medicine can treat grows steadily wider.

As they line up in emergency rooms, the health-careless will never know what they're missing. But the health-conscious will find out that they are paying for yesterday's medicine, which they don't need anymore, and not getting tomorrow's, which they do. Then, inevitably, they will look for coverage tailored to their own behavior and needs.

If they were allowed to, private insurers would respond with policies openly tailored to molecular profiles and priced accordingly. Insurers already do quite a lot of that kind of tailoring indirectly, by letting employment segment and stratify insurance pools—charging less for white-collar desk policies, for example, than for blue-collar heavy lifting. Any private insurer that fails to push this kind of segmentation as far as it can will end up covering all the heart attacks, while its competitors underwrite the low-fat or high-Lipitor diets.

————————

GOVERNMENTS DON'T FACE the risks of competition, so they can insure as indiscriminately as voting taxpayers will allow. Or to similar effect, they can—and do—require private insurers to sell only one-size-fits-all policies at one-size-fits-all prices. Even before 2010, private insurers were often barred from openly tying coverage to personal chemistry by a slew of laws mandating equal treatment and barring discrimination. In 2010, Washington codified plans to send private insurers on a forced march down a road that will, for all practical purposes, transform them into Washington's sales agents and bookkeepers, writing and administering policies exactly as directed. But however it's packaged and peddled, universal health insurance requires steadfast public support—and the political center just won't hold.

First, you have the pedestrian problem of insatiable demand that, on at least one side of the divide, seems to promise costs that rise forever. The passive, clueless, and feckless must get ruinously expensive, last-ditch care because they don't show up until it's too late for anything else, and universal means what it says. The informed and engaged will stay healthy enough to demand better hair, skin, and sex along with their Lipitor; next they will demand the look-ahead drugs to head off the more complex and often chronic diseases that arrive later in life; and then finally they'll want antidotes to aging itself. And while this is less frantic, desperate medicine, with a quite different price tag, it entails the enormous up-front costs of developing new drugs for use by the seemingly healthy. If the United States doesn't continue to lead the world in making these drugs available, some other country will, and well-informed Americans will demand access to these drugs, too.

Then there's the merciless fact of global competition in other markets. The cost of health care has a big impact on labor productivity, the cost of labor, and the marginal tax rate. If California defies the new medicine's economics by requiring insurers to ignore everything but age and geography, firms can flee to Texas. If Washington sets the standard for both states, the flight will be to Bangalore or Shanghai. Efficient labor markets require efficient health insurance, which will be found only where actuaries are permitted to find out as much as the rest of us can, and craft policies accordingly.

A third, deeper problem is (depending on your politics) either base selfishness or common sense. The pocketbook-healthy eventually tired of paying for welfare that persistently failed to end poverty; the health-healthy will tire of paying for health care that persistently fails to improve health. However selfless and generous people may be, responsible types eventually despair of trying to cure self-destructive behavior from a distance.

Finally, the new medicine is too hot for even the political right to handle, and the left can hardly even acknowledge what it's all about. To pick just one politically insoluble example among many, scientists have already isolated chemical and genetic links to mental retardation. In due course, they will develop drugs to improve or compensate for genes that help shape intelligence. Then someone will have to decide whether mental acuity, say, is as important as cystic fibrosis, and if so, where insurers must set the IQ cutoff for coverage. Every such arbitrary, line-drawing exercise will leave a new group of angry people on the wrong side of the line.

———————

ALL OF US depend on the same short list of basic nutrients—a few dozen vitamins, minerals, and amino acids, along with a ration of raw calories. Cherry Garcia alone, as it happens, pretty much covers it. This simplicity made deficiency diseases quite easy to beat, once science revealed what was missing—so easy that people are sometimes surprised to learn that vitamin science earned a fistful of Nobel Prizes in its day, and that vitamins were very expensive before they became so cheap.

In the early 1920s Quaker Oats offered $900,000 for the right to use a newly developed method to enrich the vitamin D content of food. Sensing an opportunity to peddle health and pleasure in a single package, cigarette and beer companies also wanted the patent. The inventor, Professor Harry Steenbock of the University of Wisconsin, opted instead to set up an independent foundation to license the technology and return the proceeds to his lab. Ten years later, the foundation had earned more than $17 million on the patents—and rickets had almost disappeared from the United States.

Much as vitamin deficiencies did back then, chronic obesity now destroys joints, breaks bones, swells body tissues, and causes heart disease. For one exhilarating decade, Pfizer made a fortune suppressing just one molecule in the long, toxic list of things we shouldn't consume but often do. Then statin patents began to expire. On June 22, 2006, Merck still owned a statin, Zocor, that earned the company more than $3 billion a year in the United States alone. The next day, the formula belonged to humanity.

Pfizer's Lipitor, though introduced later, had quickly eclipsed Zocor in the market, and its patent still had five years to run. But Zocor was now set to take a Pyrrhic revenge. U.S. insurers immediately began jiggering co-pay schedules to migrate patients from Lipitor to generic versions of Zocor. *Consumer Reports* estimated that a wholesale shift would save $7 billion to $11 billion a year. Lipitor's market share dropped 4 percentage points in the first half of 2006. Wall Street saw it coming: a few months earlier, Pfizer's stock price had hit an eight-year low, almost 50 percent below its Lipitor peak. Zocor still sold at about $3 a pill in early 2006; generic copies were soon selling for under 50¢.

Statins end up very cheap for much the same reason that cholesterol did: there are huge economies of scale in farming cows for milk and fungi for statins, or in brewing up synthetic versions of almost anything. The way things work, $3 statins in New York in 1996 get 30¢ statins to London in 2006 and 3¢ statins to Kuala Lumpur some years later. Or much sooner than that, if drug companies are able to maintain tiered pricing schemes.

Drug companies are quite smart enough not to pursue $3 pills for 3¢ buyers, and it takes a delicate choreography of intellectual property rights and cutthroat competition to get the innovation first and the rock-bottom prices later. But governments are impatient, especially when they have promised to supply what they can't afford but can easily seize, and they are often willing to depress the value of intellectual property rights and flatten prices. All such schemes are ice cream today, and never mind about tomorrow.

———————

THAT IS THE real crisis in health care—not medicine that's too expensive for the poor but medicine that's too expensive for the rich, too expensive ever to get to market at all. Humanity is still waiting for countless more Lipitors to treat incurable cancers, Alzheimer's, arthritis, cystic fibrosis, multiple sclerosis, Parkinson's, and a heartbreakingly long list of other dreadful but less common afflictions. Each new billion-dollar pill will be delivered—if at all—by the lure of multibillion-dollar gains. We get the 3¢ pills to the poor by selling $3 pills to the rich.

With $20 trillion or so under management, Wall Street could easily double the trillion or two it currently has invested in molecular medicine. The fastest way for Washington to deliver more health, more cheaply, to more people is to reform the drug licensing process and unleash that capital by affirming the intellectual property rights needed to maintain the efficient pricing of molecular medicine.

On the other side of the pill, molecular medicine can only be propelled by the informed, disciplined consumer. Most schemes to weaken the consumer's role will end up doing more harm than good. Vague promises of one-size-fits-all universal care maintain the illusion that the authorities will take good care of everyone, and they reaffirm the obsolete and false view that patients and their doctors always know less than Washington's crowd doctors and can't find their own way to anything better than Washington's crowd-based medical scripts.

Neither Pfizer nor Washington can ever stuff health itself into a one-price, uniform, one-America box—not when health is as personal as ice cream, genes, and pregnancy, not when every mother controls her personal consumption of carbs, cholesterol, Flintstones, and Lipitor. The thought that government authority can get more bodies in better chemical balance than free markets and free people can is more preposterous than anything found in *Das Kapital*. Freedom is now pursuing a pharmacopoeia as varied, ingenious, complex, flexible, and fecund as life itself, and the pursuit will continue for as long as microbes mutate, lifestyles change, and marriages mix and match. Given time, efficient markets will deliver a glut of cheap Lipitor for every glut of cheap cholesterol. And given time, free people will find their way to a better mix.

20

THE CULTURE OF LIFE

BAD POLICIES BECOME disasters when overtaken by events. A peace-in-our-time narcotic stupefied democracies for years while Hitler seized power and built panzers. Humanity is now half a century or so into another self-induced daze that will end in another great spasm of death. The germs are on the rise again. They have also contrived, of late, to get human sociopaths to add thought and order to their random searches for new ways to devour us.

Bacteria and viruses, we have recently discovered, don't just mutate; they exchange and remix their genes, much as we do when we procreate, and they do it much faster. The 2009 swine flu happened to be a rare, four-way remix of two pig strains, one bird strain, and one human strain. It didn't prove as lethal as first feared, but the next random rewrite of this distinctly new genetic code might well be. Even if we somehow manage to eradicate HIV, some other retrovirus, equally nimble and nasty, will pick up the torch. Several are almost certainly out there already, spreading quietly and diversifying as they prepare to surface in 2030.

Deliberate remixing by people can be far deadlier than random remixing by pigs. In a study conducted to obtain a better understanding of the risk that the flu virus might do so spontaneously, Dutch researchers (funded by Washington) recently induced five mutations in two key genes to create a highly contagious form of the lethal H5N1 bird flu strain that had surfaced in 1996. In Washington and elsewhere, authorities scrambled to block publication of the paper that described their work. There's a serious risk of theft of biological weapons already on ice in old Soviet facilities and U.S. labs. The biotech facilities now being built by Malaysia, Cuba, and at least a dozen other developing countries could readily be used to brew new ones. Germs engineered to track ethnic, racial, and geographic lines—a possibility anticipated by novelist Frank Herbert in *The White Plague*—are now quite plausible. As Bill Joy, the founder of Sun Microsystems, has observed, we confront here "a surprising and terrible empowerment of extreme individuals."

Wherever they're spawned, the new germs will spread quickly. Within a decade of reaching Kinshasa, Congo's capital city, HIV had made it to every continent and almost every country on the planet. A drug-resistant pneumonia that surfaced in Spain some years ago spread to the United States, Korea, and South Africa within five years. The hardy new form of the *E. coli* bacterium that caused outbreaks of often lethal food poisoning in Europe in May 2011 was traced to seeds cultivated in Egypt.

Until the germs take charge again, our own biochemistry will continue to kill us. Multicellular life's long flight from infectious germs has left the human species with bodies that beat germs by imitating them—we exchange and remix genes when we procreate, and our bodies make certain types of random molecular errors every day, as our tissues and immune systems spawn new proteins and cells to fight off interlopers. These processes, unfortunately, also have the capacity to spawn an endless variety of biochemically distinct cancers and autoimmune, neurological, and other complex diseases. Death is apparently the price we must pay for being so much more than germs. Bacteria don't undergo "replicative aging" and can form biologically immortal colonies. Human cells are programmed to die.

———

THAT WE HAVE recently had serious trouble dealing with something as familiar as the flu should horrify us. A new vaccine has to be developed at least once a year to keep pace with the fast-mutating virus, but that process is now technically routine. Ramping up mass production, however, hinges on skill and some degree of luck. In 2004, just three companies controlled most of the skill, and one of them got unlucky. The United States was left unable to provide even the seasonal flu shots needed by the elderly and health care workers, and thus would have been helpless against a pandemic had one materialized.

HIV and the flu virus exploit the simplest kind of biochemical complexity that can thwart magic-bullet antidotes. Washington is rigid and slow; the new germs are flexible and fast. Washington is paralyzed by the fear that error is politically lethal; the new germs make genetic error—constant mutation—the key to their survival. Unable to afford the future, Washington's paymasters are systematically biased in favor of the drug licensed decades ago. Their principal concern is who should pay how much for the new, patented drug, or whether the old, cheaper generic might do as well. The whole edifice leans sharply toward the past. Germs are always future, always reinventing themselves in their ingeniously stupid way. They don't have to be smarter than our scientists anymore, just faster than our lawyers.

For similar reasons we are failing to catch up with many of the complex diseases rooted in our own chemistry. They resist assaults by our drugs by exploiting the nimble biochemically complex capacity that human bodies developed to resist assaults by germs. These diseases don't have to be smarter than our scientists, either—they will persist by remaining too diverse and complex to be subdued by treatment regimens that can make it through the crowd-science testing and screening protocols administered by the FDA and Washington's paymasters.

To continue to advance and spread its benefits more widely and equitably, molecular medicine must mimic life in all its nimble, diverse, discriminatory, and changeable complexity. It must search for drugs in search of diseases, and find ways to get them into the hands that might need them before anyone has resolved precisely how to use them effectively, because that's how retroviruses, cancer cells, and other snippets of life inside our bodies are constantly searching for new ways to sicken and kill us.

Down there, life is a community of logic, its molecular citizens neither good nor bad, just smart. They propagate to create more life and also more death. The fragments of life that choreograph all that we crave and love thrive only by staying ahead of the fragments that terrify and ultimately destroy us all. This gloomy, Hobbesian view of things may be unwelcome, but the facts are as old as the code of life. Molecules don't care how we feel about anything. Such distinctions are made only by human minds and culture.

We now have in hand the power to decode all this amoral molecular logic, and we are fast acquiring the power to control it. Germs and our own bodies will assuredly use it for both good and evil, as they always have. History teaches that human societies will, too. The power can't be contained. It will be used by inquisitive, creative, cooperative societies that embrace the culture of life—and also by societies that could never have acquired this power on their own but harbor small groups of skilled scientists who will use it to destroy. Whatever cholera and smallpox could do, human cultures of death can now do worse. The culture of life will survive and thrive only by remaining more nimble and adaptable than they are.

————

THE SCIENTIFIC AND entrepreneurial culture that launched humanity's systematic assault on hostile biochemistry emerged in a handful of countries in Western Europe during the century that spanned the lives of Jenner, Pasteur, and Ehrlich. Europe then led the way in developing the science of vaccines and antibiotics. Private money systematized their mass production, even as public authorities gradually took charge of their distribution.

After World War II, much of the science and private investment moved to the United States, and much of the biochemical intellect from around the planet followed it. In the years during which all the key tools of modern molecular medicine came of age, America (including subsidiaries of foreign companies based in America) has conducted most of the basic research, developed more genuinely new drugs than any other country, and sold them to far more patients. And wherever the drug was developed, almost all of the rest of the world's use of drugs has been heavily subsidized by the American patients who paid more for new drugs because they didn't buy them through a government-established monopoly.

The European pioneers thus came to rely on the shelter of America's pharmaceutical umbrella. Most of the rest of the world has, in turn, been content to rely on America and, to a lesser extent, Europe, for almost all of its molecular medicine, with much of it supplied by charity or piracy. The U.S. umbrella alone has allowed countries from Belgium to Burundi to save a great deal of money. And that fact alone makes nonsense of the common assertion that other countries provide better health for less money by managing it all from the top. Western Europe also spent far less protecting itself from the Soviet menace during the Cold War, and was never attacked, either—but not because it managed its defenses so much more efficiently.

Drug companies have good reason to complain, and deserve much better protection of their intellectual property worldwide. U.S. taxpayers and patients should be complaining, too. By failing to pay its share of what it costs to develop the new drugs, the rest of the world curtails private investment in their development. But the free riding isn't going to end—the know-how genie will always find ways to escape from the legal bottle. With digital and biochemical technologies, and high-tech innovation in general, Belgium, Burundi, and most of the rest will continue, in the words of Nobel economist Edmund Phelps, to "sail in the slipstream of a handful of economies that do the preponderance of the world's innovating." In this regard, as the late Harvard economist Zvi Griliches saw it, the Europeans "are so smart." And we Americans are chumps. What we need here is a stiff new dose of European culture. We should see to it that Americans get to buy drugs as cheaply as they do in Belgium, if not Burundi—or so quite a few influential Americans have concluded.

They couldn't be more wrong. Whatever America does, other countries will find ways to develop and exploit the power that biochemical technology now offers. The raw materials are cheap and the technology has already gone global. Sooner or later, other countries will find the resources, attract the human talent, accept the learn-as-you-go imperatives of a science anchored in human biochemistry, and shoulder the high front-end costs and risks that always attend the development of great new power to control the forces of nature. These countries will lead the world in developing new biochemical code and capturing the enormous health and economic benefits that it will provide.

America's main advantage today is that we have a solid lead. Selling the drugs we make worldwide, even at cut-rate prices, allows us to spread their development costs more broadly, and the benefits ripple through other sectors of our economy. A 2011 study, commissioned by Pharma but conducted by the independent Battelle technology group, estimates that the industry's payrolls, supply chains, and innovation add almost $1 trillion a year to the U.S. GDP. There are other global benefits, too. American vaccines and drugs have saved countless lives, indiscriminately, wherever people dwell on earth. The AIDS relief program launched by George W. Bush earned America the deep respect and enduring gratitude of many of the poorest of the poor in Africa.

Over the long term, the power to control the code of life is the power to maintain and control every aspect of vigorous, resilient, productive civilization. Here, as elsewhere, some societies will be eager to take control of their own destiny, and others will be content to sail in the slipstream, shun the power of science altogether, or enlist it only in pursuit of destructive ends. That cultural schism is certainly found within America itself—we have room enough for every culture. And it is found, likewise, among the member states of the European Union.

In one key respect, however, America is uniquely qualified to serve as the cradle of modern pharmacology. Spend an hour or two in New York's Times Square and you will brush shoulders with almost every gene and lifestyle found on the face of the earth. Most every infectious germ will eventually get there, too. Whether America will continue to lead the world in decoding and learning to control the biosphere remains to be seen.

———————

IN KNOWLEDGE-BASED industries, as in life, constant change is the key to survival. Margins always collapse when yesterday's technology matures, key patents expire, manufacturing processes get standardized, and the next-generation technology emerges. The digital economy discovered the Darwinian economic law years ago: innovate or die. The law applies to drugs, too: tomorrow's profits always depend on taming the next shard of endlessly variable human chemistry, or beating nature's next new pathogen.

In a welcoming environment, Wall Street, venture capitalists, monster drug companies, small biotechs, research hospitals, and many others would be pouring intellect and money into this process. The biosphere offers unlimited opportunity for valuable innovation. The technology is new, fantastically powerful, and constantly improving; the demand for what it can supply is insatiable. While drug companies have made fortunes selling one-size-fits-all cures for very common problems, over the long term there is far more money to be made in tracking unhealthy differences down to their fragmented molecular roots. Drugs that target the flaws in our own chemistry must often become an integral part of daily life, and thus can end up extremely profitable even when they address problems that aren't very common at all. Wall Street adores many of those that have been found so far.

The money is equally happy to develop the sniffers that expose biochemical diversity and take drug markets apart. "Theranostics," the combination of drugs with patient-screening kits, makes sense because greatly improving the product's performance at modest cost almost always makes sense in a competitive market. And if Pfizer doesn't develop the kit that tells patients that they would be better off using someone else's drug, Lilly or Roche will take care of it.

If we let them, drug markets will strive endlessly to do exactly what markets do best, and what drug markets most need. Propelled as they are by dispersed initiative and private choice, free markets are uniquely good at extracting and synthesizing information that's widely dispersed among innovators, investors, workers, and customers—"personal knowledge," in the words of Michael Polanyi, the brilliant Hungarian British chemist, economist, and philosopher. In no other market are the sellers so dependent on information that is in the sole possession of the buyers. Nowhere could the free market's information-extracting genius be more important and valuable than in a market for products whose value depends on their ability to mirror biochemical information inside the people who use them.

The reactionary fear that market forces will be brutally Darwinian in deciding whom to cure overlooks the inherently social forces that rule down at the molecular level of life where drugs operate. The targeted drugs that will emerge from the free-market pursuit of life will be fiercely discriminatory, but also not discriminatory at all. Saquinavir doesn't care a fig about sexual orientation; it hates HIV protease and nothing else. And if—as now appears possible—other protease inhibitors, each slightly different, hate herpes, the common cold virus, or a key link in the chemistry of osteoporosis, strokes, or Alzheimer's, the hatred won't hinge on standard forms of bigotry. BiDil, a drug that the FDA licensed in 2005 for use by "self-identified black patients," didn't sell and was withdrawn three years after it was licensed. Its backers blamed racial bias in the health care system. The next BiDil will be for people who match a color-blind biochemical profile diagnosed by a dipstick.

The future will also end up economically indiscriminate, because copying know-how is so cheap. People with disorders not yet covered by good drugs are victims of our collective ignorance, not their individual poverty. New knowledge, the most valuable ingredient of every drug, ends up shared for free with everyone when intellectual property rights expire. The cures in the code are the ones that make health care equitable, affordable, and accessible to all. Public policy should be framed accordingly.

No clinical trial can prove that researchers, drug companies, doctors, and patients should have the right to communicate and collaborate more freely, that academics should be permitted to patent drugs developed under federal grants (as a Reagan-era law allows), that drug-related intellectual property rights in general should be strengthened, that more drugs should be declared orphans under the Orphan Drug Act, that adaptive trials are better, that more drugs should receive accelerated approval, that doctors and patients should have more discretion and control instead of less, and that we should welcome biochemical sniffers and digital networks that help doctors and patients decide for themselves what they might need and then mobilize to fight for it.

Nor can it be denied that this agenda has an ideological slant. It favors dispersion of information, authority, and economic interest. It relies less on electing drugs in large biochemical referenda overseen by Washington, and more on town meetings convened by biochemists and doctors. It benefits those who add their own intelligence to the drug's when they use it, and it may endanger those who don't. It requires parents to pay more for new pills today in order to get cheap generics to their grown children and new and better drugs to their grandchildren. It accepts that even while still under patent, medicine brewed by the vat with Wall Street's money provides far more health care, far more cheaply, than any alternative.

In support of this agenda we can, however, invoke the biochemical logic of drugs and patients. The patient's chemistry matters as much as the drug's. Americans are biochemically diverse. The intractable diseases that we now confront are biochemically complex. Trying to work out all the science up front in a handful of conventional FDA-approved clinical trials leads to quagmire and stifles the most important part of the learning before it begins. We will need a broad range of different drugs to deal with the biochemical diversity that underlies many of our diseases. We won't get them if we let the paymasters stage further rounds of FDA-like trials that aspire to lead us all to the cheapest, one-size-fits-all cures.

That said, the best that free markets can do is to keep stocking the pharmacy with an ever-expanding array of drugs as diverse as the human and microbial chemistry that shapes our health. Freedom gives people the freedom to choose, but not always the wisdom to choose well. No free society will ever manage to get the right drugs into every person whose health would improve by taking them, or ensure that every drug is used exclusively to improve health. So long as we remain free, Americans will, in matters of health, live in the land of *ex uno plura*.

VIEWED FROM THE perspective of Washington's paymasters, the three most insidious diseases in today's America are diversity, freedom, and privacy—the diversity embedded in our genes and lived in our freewheeling private life-styles, which together shape our health. For the rest of us, the insidious peril is the conviction that health rooted in the diversity and freedom of three hundred million Americans should be managed by small committees convened by federal authorities.

The culture of life compassionately respects the individual's right to choose the hospice and a peaceful passing in the company of loved ones rather than doctors—a right, one might add, that many wise doctors choose to exercise themselves. But it also accepts and honors the impulse to look ahead, start trying to intercept molecular problems long before they become clinical problems, and rage rather than go gentle into the night, because these impulses are inseparable from a way of living that is optimistic, self-reliant, willing to accept responsibility, and eager to take charge, inseparable from the correct and supremely important belief that private choices, intelligently made, are the most powerful medicine of all.

So long as patients, doctors, biochemists, and private money decide when enough is enough, medicine will never stop searching for—and finding—new drugs to intercept diseases early, slow aging, and postpone death. And ordinary people will never stop craving more and better, because there is no limit to human hunger for vigor, beauty, intelligence, and more years of active life. America can afford the culture of life because it is also the culture of prosperity.

IN JUNE 2011 Britain launched a clinical trial to test a monoclonal antibody drug harvested from a genetically engineered tobacco plant and incorporated in a cream that prevents HIV from colonizing the vaginal lining. Thus, a plant that has probably killed more people in a century than small-pox killed in a millennium may soon be enlisted to save humanity from a virus that is currently infecting five new people for every two who start treatment.

Meanwhile, others were learning how to turn the murderous virus itself, killer of about thirty-five million people so far, into a magical cure. In April 2012 oncologists at the Children's Hospital of Philadelphia used a genetically modified form of HIV to insert new genes into healthy immune-system T cells collected from six-year-old Emily Whitehead. She had been fighting leukemia for two years and was on the brink of death. The new genes allowed Emily's healthy cells to recognize and attack her immune system's B cells, which harbored the cancerous cells. An overproduction of a cytokine triggered by the treatment nearly killed her but was suppressed when one of her doctors recalled that an arthritis drug targets the cytokine in question. The staff in the intensive care unit sang "Happy Birthday" when she awoke from a coma on the day she turned seven. Seven months later Emily was back in school, playing soccer, and walking her dog, Lucy. The *New York Times* ran her story shortly before Christmas.

Our ability to use life itself to manufacture and deliver the code that cures is the one area where our mastery of the code of life has already over-taken nature's. Life is fecund: it can't be contained, it replicates fast, and it spreads indiscriminately. But we now know how to replicate and distribute at least some snippets of lifesaving biochemical code even faster. We proved it when we beat smallpox by spreading the vaccine for it faster than the virus could spread from person to person. Since then, we have learned to read biochemical code, distribute it worldwide at the speed of light, and enlist more capital and human intellect in its design and manufacture. If we choose to, we can cultivate enough bacteria to manufacture human insu-lin faster than all the diabetics on the planet can use it. We can churn out monoclonal antibodies faster than all the cancerous breasts on earth could spawn cancer cells. And we can cultivate enough tobacco and crippled HIV to save more lives than they ever destroyed. Such is the redemptive power of the culture of life.

When the wealthy spend lavishly trying to beat the unbeatable, we should thank them. We should not only permit but encourage afflu-ent people to spend more, on a much broader range of drugs, including leading-edge experimental treatments, than government paymasters could ever afford or include in any one-size-fits-all health care scheme. We need the very rich here—and, happily, they need the rest of us even more. They are built out of the same molecules as we are. They can't lock away the know-how that their bodies and their money help develop, and they can't develop it on their own, either. They need the rest of us to help build the databases that reveal how all the molecular pieces fit together.

Anchored in the sharing of biochemical knowledge, the new socialized medicine offers far more power and is far more egalitarian than anything imagined during the days when much of the developed world was socializ-ing hospital beds. Knowledge is inherently, unstoppably socialist: it spreads wealth faster than autocrats can spread poverty, and it can spread life faster than germs can spread death. The orderly socialization of potent know-how is also the one form of socialism that can be harmonized with free markets and vigorous competition. The sharing of knowledge doesn't cen-tralize power; it disperses it. With this unique form of wealth, the hard part isn't the sharing; it's finding ways to limit the sharing long enough to give markets the incentive to develop the know-how in the first place. With germ-killing know-how, there's the further problem that indiscriminate sharing quickly breeds drug-resistant germs. These problems are solvable, but not by people who reflexively assume that more free sharing sooner is always better.

Recent political developments notwithstanding, there is reason to remain optimistic about the role that America will continue to play as the biochemical and digital revolutions converge. By democratizing the power to read biochemical text, the sniffers are rapidly exposing the limits and frailties of the old medical media, beginning with the government licenses, labels, mandates, and proscriptions that attempt to dictate who may sell and who may use what kind of molecular medicine for what purposes. By pooling and distributing what the sniffers detect, the digital cloud will democratize our understanding of health and disease. The cloud will end up linking the great repository of the old code, patients themselves, with the developers of the new, the biochemists and doctors who design clever fixes and patches and find new ways to use them well. If Washington persists in forbidding or strictly controlling such connections, the information will move through intermediaries, based in the United States or abroad. As biochemical knowledge accumulates and its implications become clear, free people will come to understand what they are missing, and insist on taking far more personal control of the sniffers and drugs that will let them control the arc of their biological lives.

While the good biochemical code will keep getting better, the bad will keep getting worse. And it, too, knows how to spread fast. The only way to stay ahead of murderous life spawned in our own bodies, in some distant rain forest, or in a bioterrorist's lab is to nurture a bold, optimistic, life-affirming culture. A culture of discovery and creativity that is willing to take risks and invest patiently in the future. One that embraces the active, endless search for a better understanding of how life works, and how the murderous parts can be controlled. A culture that welcomes the fact that its adherents share an insatiable hunger for life, and will, as they grow more wealthy, devote more of their wealth to preserving it.

Ars moriendi, ars vivendi est. That will never change. But we now have the capacity to understand the art of dying well enough to transform our reverence for life into temporal power that will deliver us from much evil and reaffirm our faith in the ultimate goodness and mercy of creation.

ACKNOWLEDGMENTS

SEVERAL CHAPTERS OF this book draw on material previously published in the Manhattan Institute's *City Journal*, *Forbes* magazine, and *Commentary*, and I am indebted to the editors of those journals for their input and assistance. Brian Anderson and Ben Plotinsky of the Manhattan Institute's *City Journal* also reviewed the book in its entirety and provided invaluable editorial input throughout. I am grateful for the further editorial assistance I received from Tim Bartlett at Basic Books. Kaitlin Keegan, Jessica Pascoal, and Yevgeniy Feyman provided excellent research assistance. I am, finally, deeply grateful for the patient, unstinting support and help I have received from the Manhattan Institute for Policy Research and its president, Larry Mone.

NOTES

WETWARE

xi   **Cancer cells, as described by Sherwin Nuland:** Sherwin B. Nuland, *How We Die: Reflections on Life's Final Chapter* (New York: Vintage Books, 1993), 208.

xii   **The Hatfields' endless feud:** "Tumors May Have Fueled Hatfield-McCoy Feud," *Vanderbilt Magazine*, Fall 2007. But for another perspective that tells the story as the Hatfields persecuting the McCoys, see Dean King, *The Feud: The Hatfields and the McCoys: The True Story* (Little Brown, 2013).

xii   **Can't sleep? It could be worse:** For an engrossing account of fatal familial insomnia and other prion diseases, read D. T. Max, *The Family That Couldn't Sleep: A Medical Mystery* (New York: Random House, 2007).

xiii   **Six related children in northern Pakistan:** Michael Hopkin, "The Mutation That Takes Away Pain," *Nature News*, December 13, 2006.

xiv   **"naive medical reductionism":** Richard Lewontin, *It Ain't Necessarily So: The Dream of the Human Genome and Other Illusions* (New York: New York Review of Books, 2000); Richard Lewontin, *The Triple Helix: Gene, Organism and Environment* (Cambridge: Harvard University Press, 2001); Rob Dorit, "Defying Genomania," *American Scientist*, September-October 2010.

xv   **It touches down in the office of the doc-in-a-box:** Julie Appleby and Sarah Varney, "Wal-Mart Plans Ambitious Expansion into Medical Care," NPR.org, November 9, 2011; Walmart Strategic Partner Request for Information, "Request for Information: Walmart Health and Wellness," http://media.npr.org/assets/blogs/health/images/2011/11/Walmarthealthpartnerships.pdf (accessed December 21, 2012).

CHAPTER 1 THE TRIUMPH—AND LIMITS—OF SOCIALIZED MEDICINE

3   **In *The Moral Imagination*: Gertrude Himmelfarb:** *The Moral Imagination: From Edmund Burke to Lionel Trilling* (Chicago: Ivan R. Dee, 2006), 67.

3   **Socialized medicine's finest hour arrived a century later:** Centers for Disease Control, "Global Effort Pays Off: Smallpox at Target 'Zero,'" *Dateline: CDC* 11, no. 10 (October 1979); Richard Preston, "The Demon in the Freezer," *New Yorker*, July 12, 1999.

4   **The measles immunization payoff was thirteen to one:** Centers for Disease Control, "Achievements in Public Health, 1990–1999: Impact of Vaccines Universally Recommended for Children—United States, 1990–1998," *Morbidity and Mortality Weekly Report* 48, no. 12 (April 2, 1999): 243–248.

4   **In 2001, every dollar spent on immunization:** Fangjun Zhou, "Economic Evaluation of Routine Childhood Immunization with DTaP, Hib, IPV, MMR and HepB Vaccines in the United States, 2001," Centers for Disease Control PowerPoint Presentation, http://archive.hhs.gov/nvpo/meetings/jun2003/zhou.ppt (accessed January 23, 2013); M. C. Lindley and A. Bhatt, "Child, Adolescent, and Adult Immunizations Evidence—Statement," National Business Group on Health, November 23, 2011, www.businessgrouphealth.org/preventive/topics/immunizations.cfm (accessed January 23, 2013).

6   **"A few members, bent upon investigating":** Quoted in Stephen Halliday, "Death and Miasma in Victorian London: An Obstinate Belief," *British Medical Journal* 323, no. 7327 (December 22, 2001): 1469–1471.

7   **In a recent survey by the *British Medical Journal*:** Annabel Ferriman, "BMJ Readers Choose the 'Sanitary Revolution' as Greatest Medical Advance Since 1840," *British Medical Journal* 334 (January 18, 2007): 111.

7   **One resident, Henning Jacobson, refused:** *Jacobson v. Massachusetts*, 197 U.S. 11, 26 (1905).

12   **"Yours is the comfortable reflection":** Quoted in David Koplow, *Smallpox: The Fight to Eradicate a Global Scourge* (Los Angeles: University of California Press, 2003), 21.

12   **Washington did its bit here, too—the national "vaccine agent":** "The North Carolina Accident," *Raleigh Register*, 1822, www.lib.unc.edu/ncc/ssgh/pox.html (accessed December 26, 2012).

CHAPTER 2 SNIFFERS

13   **In early 2012, scientists at Stanford University described:** Rui Chen et al., "Personal Omics Profiling Reveals Dynamic Molecular and Medical Phenotypes," *Cell* 148, no. 6 (March 16, 2012): 1293–1307; Allison Blass, "Genetics in Action: Researcher Watches His Own Body Develop Diabetes," *Diabetes Mine*, April 26, 2012, www.diabetesmine.com/2012/04/genetics-in-action-researcher-watches-his-own-body-develop-diabetes.html (accessed December 19, 2012); Krista Conger, "Revolution in Personalized Medicine: First-Ever Integrative 'Omics' Profile Lets Scientist Discover, Track His Diabetes Onset," Stanford School of Medicine, March 15, 2012, http://med.stanford.edu/ism/2012/march/snyder.html (accessed December 19, 2012).

15   **Sensor chemicals on the surface:** "Colorimetric Sensor Array," Metab-
        olomx, http://isensesystems.com/technologies/sensor (accessed December
        19, 2012); "Applications: Lung Cancer Detection," Metabolomx, http://
        metabolomx.com/applications/lung-cancer-detection (accessed January 22,
        2013); "Applications: Tuberculosis," Metabolomx, http://metabolomx.com
        /applications/tuberculosis (accessed January 22, 2013).

15   **He then watched his cholesterol level drop:** Conger, "Revolution in Per-
        sonalized Medicine."

17   **Researchers at Massachusetts General Hospital:** Catherine Alix-Panabières
        and Klaus Pantel, "Circulating Tumor Cells: Liquid Biopsy of Cancer," *Clini-
        cal Chemistry* 59, no. 1 (January 2013): 110–118.

17   **Breast cancer, one form of diabetes:** Susan Gaidos, "A Spitting Image of
        Health: How Saliva Can Help Doctors Diagnose Disease," *Science News*
        180, no. 11 (November 19, 2011): 26.

17   **One website has coordinated a social-network-based online trial:** Amy
        Dockser Marcus, "The Future of Social Network-Based Trials," *Wall Street
        Journal*, April 25, 2011, http://blogs.wsj.com/health/2011/04/25/the-future-of
        -social-network-based-trials (accessed January 22, 2013).

18   **The introduction of home pregnancy testing kits:** Sarah A. Leavitt, "'A
        Private Little Revolution': The Home Pregnancy Test in American Cul-
        ture," *Bulletin of the History of Medicine* 80, no. 2 (Summer 2006): 317–345.

19   **As three doctors working with the VA Outcomes Group:** H. Gilbert
        Welch, Lisa Schwartz, and Steven Woloshin, "What's Making Us Sick Is
        an Epidemic of Diagnoses," *New York Times*, January 2, 2007.

21   **Walgreens was poised to start selling:** U.S. Food and Drug Administra-
        tion, "Letter to Pathway Genomics Corporation Concerning the Pathway
        Genomics Genetic Health Report," May 10, 2010, www.fda.gov/Medical
        Devices/ResourcesforYou/Industry/ucm211866.htm (accessed December 19,
        2012); Andrew Pollack, "Walgreens Delays Selling Personal Genetic Test
        Kit," *New York Times*, May 12, 2010; Jeffrey Shuren, "Direct-to-Consumer
        Genetic Testing and the Consequences to the Public," statement before
        the Subcommittee on Oversight and Investigations, Committee on Energy
        and Commerce, and U.S. House of Representatives, July 22, 2010, www.
        fda.gov/NewsEvents/Testimony/ucm219925.htm (accessed December 19,
        2012).

21   **Invoking its authority to license every diagnostic "contrivance":** Fed-
        eral Food, Drug, and Cosmetic Act § 201(*h*), 21 *U.S.C.* § 321(*h*); Shuren,
        "Direct-to-Consumer Genetic Testing and the Consequences to the Public."

21   **Depending on where he or she is delivered:** Amy Dockser Marcus, "New
        Tests for Newborns, and Dilemmas for Parents," *Wall Street Journal*, July 26,
        2011.

21   **The FDA is determined to make sure, however:** Letter from Alberto
        Gutierrez, Office of In Vitro Diagnostic Evaluation and Safety for the Cen-
        ter for Device and Radiological Health, to Anne Wojcicki, 23andMe Per-
        sonal Genome Service, "RE: 23andMe Personal Genome Service," June 10,

2010, www.fda.gov/downloads/MedicalDevices/ResourcesforYou/Industry/UCM215240.pdf (accessed January 23, 2013).

CHAPTER 3 INTELLIGENT DESIGN

23   **The drug—a monoclonal antibody (mAb) designed:** See Robert Bazell, *Her-2: The Making of Herceptin, a Revolutionary Treatment for Breast Cancer* (New York: Random House, 1998).

26   **The drug, however, turned the skin of both children:** Morton A. Meyers, *Happy Accidents: Serendipity in Major Medical Breakthroughs in the Twentieth Century* (New York: Arcade, 2011), 51–52.

28   **The structure of the real target:** Jason Thomas Maynes, "The Role of Protein Structure in Rational Drug Design and Pharmaceutical Development," *University of Alberta Health Sciences Journal* 1, no. 1 (October 2004): 7–8.

29   **The computers can't yet do it all:** Seth Cooper et al., "Predicting Protein Structures with a Multiplayer Online Game," *Nature* 466, no. 7307 (August 5, 2010): 756–760; Rachel Ehrenberg, "World of Proteincraft," *Science News* 178, no. 5 (August 28, 2010): 7.

30   **Selling mAbs now earns drug companies tens of billions:** Pablo A. Scolnik, "mAbs: A Business Perspective," *mAbs* 1, no. 2 (March-April 2009): 179–184.

30   **And as of 2009, more than twenty clinical trials:** Christine Klinguer-Hamour, Veronique Caussanel, and Alain Beck, "Monoclonal Antibodies for Treating Infectious Diseases," *Medecine/Sciences (Paris)* 25, no. 12 (December 2009): 1116–1120.

30   **Gleevec is "proof that molecular targeting works":** "FDA Approves Important New Leukemia Drug," National Cancer Institute at the National Institutes of Health, May 10, 2001, www.cancer.gov/newscenter/pressreleases/2001/gleevecpressrelease (accessed December 27, 2012).

31   **In his 2003 book *Magic Cancer Bullet*:** Daniel Vasella, *Magic Cancer Bullet: How a Tiny Orange Pill May Rewrite Medical History* (New York: HarperCollins, 2003).

CHAPTER 4 COMING APART

35   **"No other class of drug has been examined":** Carl Djerassi, "The Pill: Emblem of Liberation," *British Medical Journal* 334 (January 4, 2007): s15.

36   **Estrogen itself, however, is used to treat other:** Nathan Seppa, "Using Estrogen to Combat Persistent Breast Cancer," *Science News* 176, no. 6 (September 12, 2009): 13.

36   **"The story of estrogen's role":** Renee Twombly, "Estrogen's Dual Nature? Studies Highlight Effects on Breast Cancer," *Journal of the National Cancer Institute* 103, no. 12 (June 22, 2011): 920–921.

36   **The first published reference to the "triple negative" form:** W. D. Foulkes et al., "Triple Negative Breast Cancer," *New England Journal of Medicine* 363 (2010): 1938–1948.

37   **"We expect the refined algorithm to change":** A. Goldhirsch et al., "Thresholds for Therapies: Highlights of the St. Gallen International Expert Consensus on the Primary Therapy of Early Breast Cancer 2009," *Annals of Oncology* 20, no. 8 (2009): 1319–1329, quoted in "Radically Different Approach to Treating Early Breast Cancer," *Science Daily,* June 19, 2009, www.sciencedaily.com/releases/2009/06/090617201804.htm (accessed January 20, 2013).

37   **Biochemists and oncologists now have in hand:** Kate Kelland, "Study Suggests Breast Cancer Is Clutch of 10 Diseases," Reuters, April 18, 2012; Christina Curtis et al., "The Genomic and Transcriptomic Architecture of 2,000 Breast Tumours Reveals Novel Subgroups," *Nature* 486 (June 21, 2012): 346–352.

37   **Some concede that there's no link:** "Is Abortion Linked to Breast Cancer?" American Cancer Society, www.cancer.org/cancer/breastcancer/more information/is-abortion-linked-to-breast-cancer (accessed January 20, 2013); National Cancer Institute, National Institutes of Health, Department of Health and Human Services, "Cancer Facts: Pregnancy and Breast Cancer Risk," December 30, 2003, www.emory.edu/KomenEd/PDF /Breast%20Cancer/Pregnancy%20&%20Breast%20Cancer.pdf (accessed January 2, 2013); "Abortion, Miscarriage, and Breast Cancer Risk," National Cancer Institute, www.cancer.gov/cancertopics/factsheet/Risk/abortion -miscarriage (accessed January 20, 2013).

38   **In 2011, a group of researchers published:** Linh M. Tran et al., "Inferring Causal Genomic Alterations in Breast Cancer Using Gene Expression Data," *BMC Systems Biology* 5, no. 121 (August 1, 2011).

40   **Researchers investigating the wild mutability:** Anath C. Lionel et al., "Rare Copy Number Variation Discovery and Cross-Disorder Comparisons Identify Risk Genes for ADHD," *Science Translational Medicine* 3, no. 95 (August 10, 2011): 1–11; Tina Hesman Saey, "Shared Differences: The Architecture of Our Genomes Is Anything but Basic," *Science News* 175, no. 9 (April 25, 2009): 16–20.

40   **In October 2012, the 1000 Genomes Project:** Tina Hesman Saey, "1000 Genomes Pilot a Hit with Geneticists," *Science News* 178, no. 11 (October 28, 2010): 14; Elizabeth Pennisi, "1000 Genomes Project Gives New Map of Genetic Diversity," *Science* 330, no. 6004 (October 29, 2010): 574–575; "Human Genome: Genomes by the Thousand," *Nature* 467 (October 27, 2010): 1026–1027.

40   **The discovery of CNVs, in the words of one geneticist:** Quoted in Saey, "1000 Genomes Pilot a Hit with Geneticists."

41   **Scientists had already used information from the study to identify:** The 1000 Genomes Project Consortium, "An Integrated Map of Genetic Variation from 1,092 Human Genomes," *Nature* 491 (November 1, 2012): 56–65; Robert Lee Hotz, "Personalized Medicine Moves Closer," *Wall Street Journal,* October 31, 2012.

41   **Another study, completed a few months earlier:** Tina Hesman Saey, "Uncommon Carriers: People Have a Surprising Number of Rare Genetic

Variants," *Science News* 182, no. 4 (August 25, 2012): 28–29; Jacob A. Tennessen et al., "Evolution and Functional Impact of Rare Coding Variation from Deep Sequencing of Human Exomes," *Science* 337, no. 6090 (July 6, 2012): 64–69; Matthew R. Nelson et al., "An Abundance of Rare Functional Variants in 202 Drug Target Genes Sequenced in 14,002 People," *Science* 337, no. 6090 (July 6, 2012): 100–104.

41   **Most of these rare variants:** Saey, "Uncommon Carriers."

41   **Because rare variations tend to cluster geographically:** Nelson et al., "An Abundance of Rare Functional Variants."

41   **So the best that drug designers can do is search:** Ibid.

41   **Seemingly common disorders, it now appears:** Ibid.; "The Copy Number Variation (CNV) Project," Wellcome Trust Sanger Institute website, www.sanger.ac.uk/research/areas/humangenetics/cnv (accessed January 3, 2013); Nicholas J. Schork et al., "Common vs. Rare Allele Hypotheses for Complex Diseases," *Current Opinion in Genetics and Development* 19, no. 3 (June 2009): 212–219; Stuart Wolpert, "'Rare' Genetic Variants Are Surprisingly Common, Life Scientists Report," UCLA News Release, May 17, 2012, http://newsroom.ucla.edu/portal/ucla/rare-genetic-variants-are -surprisingly-234074.aspx?link_page_rss=234074 (accessed January 3, 2013); "Common Copy Number Variations Unlikely to Contribute Significantly Toward Common Diseases," e! Science News, March 31, 2010, http:// esciencenews.com/articles/2010/03/31/common.copy.number.variations .unlikely.contribute.significantly.toward.common.diseases (accessed January 3, 2013).

44   **A predictive engine developed from a high-powered statistical analysis:** Rachel Badovinac Ramoni et al., "Predictive Genomics of Cardioembolic Stroke," *Stroke* 40, Supplement no. 3 (March 2009): S67–S70; "Predicting Risk of Stroke from One's Genetic Blueprint," Boston Children's Hospital news release, February 25, 2009, http://childrenshospital.org/newsroom/ Site1339/mainpageS1339P510.html (accessed January 6, 2013).

44   **Another model provides even higher accuracy in predicting:** Paola Sebastian et al., "Genetic Dissection and Prognostic Modeling of Overt Stroke in Sickle Cell Anemia," *Nature Genetics* 37, no. 4 (April 2005): 435–440.

CHAPTER 5 REASSEMBLING THE PIECES: PART 1

49   **As geneticist Steve Jones puts it:** Steve Jones, *Darwin's Ghost: The Origin of Species Updated* (New York: Ballantine, 2000), 292.

51   **In a 2012 paper, "Quantifying the Chemical Beauty of Drugs":** G. Richard Bickerton et al., "Quantifying the Chemical Beauty of Drugs," *Nature Chemistry* 4 (January 24, 2012): 90–98; Rachel Ehrenberg, "Measuring What Makes a Medicine," *Science News* 181, no. 5 (March 10, 2012): 18.

53   **"We're moving from a paradigm of detection":** Quoted in Rachel Ehrenberg, "Network Analysis Predicts Drug Side Effects," *Science News* 181, no. 2 (January 28, 2012): 12; Aurel Cami et al., "Predicting Adverse Drug

Events Using Pharmacological Network Models," *Science Translational Medicine* 3, no. 114 (December 21, 2011): 114ra127.

53 **By mining ten years' worth of clinicians' notes:** Francisco S. Roque et al., "Using Electronic Patient Records to Discover Disease Correlations and Stratify Patient Cohorts," *PLoS Computational Biology* 7, no. 8 (2001): e1002141; Rachel Ehrenberg, "Mining Electronic Records Yields Connections Between Diseases," *Science News* 180, no. 8 (October 8, 2011): 16.

53 **Yet another team developed what one member describes:** Joel T. Dudley et al., "Computational Repositioning of the Anticonvulsant Topiramate for Inflammatory Bowel Disease," *Science Translational Medicine* 3, no. 96 (August 17, 2011): 96ra76; Marina Sirota et al., "Discovery and Preclinical Validation of Drug Indications Using Compendia of Public Gene Expression Data," *Science Translational Medicine* 3, no. 96 (August 17, 2011): 96ra77; Amy Dockser Marcus, "Researchers Show Gains in Finding Reusable Drugs," *Wall Street Journal*, August 18, 2011.

55 **In June 2012 came the announcement:** The Human Microbiome Project Consortium, "Structure, Function and Diversity of the Healthy Human Microbiome," *Nature* 486, no. 7402 (June 14, 2012): 207–214; Ron Winslow and Jonathan D. Rockoff, "Gene Map of Body's Microbes Is New Health Tool," *Wall Street Journal*, June 14, 2012.

55 **Other microbiologists, according to one recent count:** Gina Kolata, "The New Generation of Microbe Hunters," *New York Times*, August 30, 2011.

55 **Each genome, in the words of Dr. David A. Relman: Ibid.,** David A. Relman, "Microbial Genomics and Infectious Diseases," *New England Journal of Medicine* 365 (July 28, 2011): 347–357.

56 **Some genes go a step further:** Jones, *Darwin's Ghost*, 92.

56 **They are working on a long-term flu vaccine:** Tina Hesman Saey, "Model for Powerful Fighters from Existing Drugs," *Science News* 177, no. 1 (January 2, 2010): 11.

57 **"Cancer immunotherapy comes of age":** Ira Mellman et al., "Cancer immunotherapy comes of age," *Nature* 480, no. 7378 (December 22, 2011): 480–489.

57 **"By vaccinating against multiple proteins":** "Mayo Clinic Researchers Building Melanoma Vaccine to Combat Skin Cancer," Mayo Clinic, March 19, 2012, www.mayoclinic.org/news2012-rst/6765.html (accessed January 8, 2013).

58 **In August 2011 a team at the University of Pennsylvania:** Walter J. Urba and Dan L. Longo, "Redirecting T Cells," *New England Journal of Medicine* 365, no. 8 (August 25, 2011): 754–757; David L. Porter et al., "Chimeric Antigen Receptor-Modified T Cells in Chronic Lymphoid Leukemia," *New England Journal of Medicine* 365 (August 25, 2011): 725–733; Robert Bazell, "New Leukemia Treatment Exceeds 'Wildest Expectations,'" NBC News, August 10, 2011, www.msnbc.msn.com/id/44090512/ns/health-cancer/t /new-leukemia-treatment-exceeds-wildest-expectations/#.UO15vq7-uSo (accessed January 9, 2013).

CHAPTER 6 THE SOCIAL CONTRACT

64   **Assisted by mass advertising made possible:** James Harvey Young, *The Medical Messiahs: A Social History of Health Quackery in Twentieth-Century America* (Princeton, NJ: Princeton University Press, 1967).

64   **The Constitution secured every citizen's right:** *Allgeyer v. Louisiana*, 165 U.S. 578 (1897), 589.

64   **J. H. Kelly was the proud inventor:** Young, *The Medical Messiahs*, 69.

64   **Supreme Court justice Rufus Wheeler Peckham:** *American School of Magnetic Healing v. McAnnulty*, 187 U.S. 94 (1902).

64   **The Federal Food and Drugs Act of 1906:** *United States v. Johnson*, 221 U.S. 488 (1911); Young, *The Medical Messiahs*, 49. (Today the act is commonly referred to as "the Pure Food and Drug Act of 1906.")

65   **"Elixir Sulfanilimide" promptly killed 105 people:** Paul A. Offit, *The Cutter Incident: How America's First Polio Vaccine Led to the Growing Vaccine Crisis* (New Haven, CT: Yale University Press, 2005), 158.

66   **In 1938, the U.S. Public Health Service tested a pertussis vaccine:** Jim Manzi, *Uncontrolled: The Surprising Payoff of Trial-and-Error for Business, Politics, and Society* (New York: Basic Books, 2012), 77; Curtis L. Meinert, *Clinical Trials: Design, Conduct, and Analysis* (New York: Oxford University Press, 1986), 93.

66   **Eight years later, British researchers conducted:** Arun Bhatt, "Evolution of Clinical Research: A History Before and Beyond James Lind," *Perspectives in Clinical Research* 1, no. 1 (2010): 6–10.

67   **"We were dickering with them":** Quoted in John Schwartz, "Debate from the '60s Echoes Today," *Washington Post*, July 17, 1998.

68   **Thus six times between 1988 and 2002:** Food and Drug Administration, "Guidance for Industry: Collection of Race and Ethnicity Data in Clinical Trials," *Clinical Medical*, September 2005, 8–11.

68   **As the FDA sheepishly acknowledged at the time:** Food and Drug Administration, "Guidance for Industry," 3.

68   **In a 1994 directive explaining how much diversity:** National Institutes of Health, "NIH Guidelines on the Inclusion of Women and Minorities as Subjects in Clinical Research," *NIH Guide* 23, no.11 (1994).

68   **The word *liberty*, Justice Peckham had explained:** *Allgeyer v. Louisiana*, 165 U.S. 578 (1897), 589.

69   **Henning Jacobson refused:** *Jacobson v. Massachusetts*, 197 U.S. 11 (1905).

71   **In four separate opinions, seven Supreme Court justices:** *Griswold v. Connecticut*, 381 U.S. 479 (1965).

71   **Eight years later Justice Douglas made that doubly clear:** *Weinberger v. Hynson, Westcott & Dunning, Inc.*, 412 U.S. 609 (1973).

72   **As that has become increasingly clear, patients have repeatedly:** See, for example, *Abigail Alliance for Better Access to Developmental Drugs v. von Eschenbach*, 495 F.3d 695 (D.C. Cir. 2007) (en banc), *cert. denied*, 552 U.S. 1159 (2008).

74 **The opinions in these cases affirm the patient's right:** *Reyes v. Wyeth Laboratory*, 498 F.2d 1264, 1276 (5th Cir. 1974).

74 **Meanwhile, other judges had decided that no amount:** *Davis v. Wyeth Laboratories, Inc.*, 399 F.2d 121 (9th Cir. 1968).

CHAPTER 7 A VIRUS LIKE US

79 **On June 5, 1981, the Centers for Disease Control reported:** Centers for Disease Control and Prevention, "*Pneumocystis* Pneumonia—Los Angeles," *Morbidity and Mortality Weekly Report* 30, no. 21 (June 5, 1981): 1–3; Centers for Disease Control and Prevention, "Kaposi's Sarcoma and Pneumocystis Pneumonia Among Homosexual Men—New York City and California," *Morbidity and Mortality Weekly Report* 30 (July 4, 1981): 306–308; Kenneth B. Hymes et al., "Kaposi's Sarcoma in Homosexual Men—A Report of Eight Cases," *Lancet* 318, no. 8247 (September 19, 1981): 598–600.

79 **By 1992, HIV was the leading killer:** Steve Jones, *Darwin's Ghost: The Origin of Species Updated* (New York: Ballantine, 2000), 6.

80 **As the National Academy of Sciences would observe in 2000:** National Academy of Science, "Disarming a Deadly Virus: Proteases and Their Inhibitors," *Beyond Discovery*, February 2000.

80 **As Sherwin Nuland observes:** Sherwin Nuland, *How We Die: Reflections on Life's Final Chapter* (New York: Random House, 1995), 164.

80 **"One in five—listen to me":** Quoted in Michael Fumento, *The Myth of Heterosexual AIDS: How a Tragedy Has Been Distorted by the Media and Partisan Politics* (New York: Regnery Gateway, 1993), 3, citing "Women Living with AIDS," *Oprah Winfrey Show*, February 18, 1987, 2.

81 **"As death approaches":** Jones, *Darwin's Ghost*, 2.

83 **Developed elsewhere by Jerome Horwitz:** Jie Jack Li, *Laughing Gas, Viagra, and Lipitor: The Human Stories Behind the Drugs We Use* (New York: Oxford University Press, 2006), 122.

83 **Then Burroughs Wellcome acquired the drug:** Gertrude B. Elion, "The Purine Path to Chemotherapy," Nobel Lecture, Physiology or Medicine, December 8, 1988. www.nobelprize.org/nobel_prizes/medicine/laureates/1988/elion-lecture.pdf (accessed January 10, 2013).

83 **The zidovudine trial had to be terminated:** Donna E. Shalala, Jane E. Henney, Janet Woodcock, and Marcia L. Trenter, "From Test Tube to Patient: Improving Health Through Human Drugs," FDA Center for Drug Evaluation and Research (September 1999): 9–11, www.canceractionnow.org/FromTestTubeToPatient.pdf (accessed January 12, 2013); Ken Flieger, "FDA Finds New Ways to Speed Treatment to Patients," FDA Consumer Special Report on New Drug Development in the United States, January 1995, www.mdadvice.com.c25.sitepreviewer.com/resources/clinical_trials/new.htm (accessed January 12, 2013).

87 **As of late 2011, the largest such engine:** EuResist, "EuResist Prediction System," http://engine.euresist.org (accessed March 10, 2012).

87 **The study was dubbed "Engine Versus Experts":** Ibid.; Steven To-masco, "IBM and EU Partners Create a Better Way to Fight AIDS Vi-rus," IBM news release, June 2, 2009, www-03.ibm.com/press/us/en/press release/27627.wss (accessed January 13, 2013). For a detailed discussion of how the EuResist Engine operates, see Michael Rosen-Zvi et al., "Selecting Anti-HIV Therapies Based on a Variety of Genomic and Clinical Factors," *Bioinformatics* 24, no. 13 (2008): i399-i406.

87 **"The EuResist system is a linear combination":** EuResist, "EuResist Pre-diction System."

88 **"Once we understand a cancer cell":** Quoted in Marilynn Marchione, "Gene-Based Lung Cancer Drug Shows Promise," NBC News, June, 5, 2010, www.msnbc.msn.com/id/37527542/ns/health-cancer/t/gene-based-lung -cancer-drug-shows-promise/#.T2nwTMUgfZc (accessed January 13, 2013).

89 **At the Massachusetts General Hospital:** Tina Hesman Saey, "Tumor Tell-All," *Science News* 180, no. 7 (September 24, 2011): 18–21.

89 **Working with the drugs they do have, oncologists:** Amy P. Abernethy et al., "Systematic Review: Reliability of Compendia Methods for Off-Label Oncology Indications," *Annals of Internal Medicine* 150, no. 5 (March 3, 2009): 336–343; "The Off-Label Use of Drugs in Oncology: A Position Paper by the European Society for Medical Oncology (ESMO)," *Annals of Oncology* 18, no. 12 (2007): 1923–1925.

89 **In early 2013, IBM announced the arrival:** "IBM Watson Hard at Work: New Breakthroughs Transform Quality Care for Patients," IBM press re-lease, February 8, 2013, www-03.ibm.com/press/us/en/pressrelease/40335. wss (accessed February 13, 2013).

90 **By early 1998 the FDA had granted accelerated approval:** "Providing Access to Promising Therapies for Seriously Ill and Dying Patients," state-ment of Michael A. Friedman, M.D., lead deputy commissioner, Food and Drug Administration, Department of Health and Human Services, before the Committee on Government Reform and Oversight, U.S. House of Representatives, April 22, 1998, www.fda.gov/NewsEvents/Testimony/ucm 115120.htm (accessed April 15, 2013).

90 **In a September 2012 report, President Obama's Council of Advisors:** President's Council of Advisors on Science and Technology, *Report to the President on Propelling Innovation In Drug Discovery, Development, and Eval-uation* (September 2012), www.whitehouse.gov/sites/default/files/microsites /ostp/pcast-fda-final.pdf (accessed April 15, 2013).

90 **Critics of accelerated approval:** John R. Johnson et al., "Accelerated Approval of Oncology Products: The Food and Drug Administration Experience," *Jour-nal of the National Cancer Institute* 103, no. 8 (April 20, 2011): 636–644.

CHAPTER 8 DRUG SCIENCE FROM THE BOTTOM UP

93 **The 1981 report that prompted Washington:** Centers for Disease Control and Prevention, "*Pneumocystis* Pneumonia—Los Angeles," *Morbidity and Mortality Weekly Report* 30, no. 21 (June 5, 1981): 1–3.

94   **But two months after it licensed AZT, the FDA:** Flieger, "FDA Finds New Ways to Speed Treatment to Patients"; Frank E. Young, "The Role of the FDA in the Effort Against AIDS," *Public Health Reports* 103, no. 3 (May-June 1988): 244–245.

95   **This cluster of overlapping and complementary rules:** Flieger, "FDA Finds New Ways to Speed Treatments to Patients"; FDA, "Treatment Use of Investigational Drugs—Information Sheet," www.fda.gov/Regulatory-Information/Guidances/ucm126495.htm (accessed April 15, 2013). For a review of how many of these rules evolved during the first two decades of the battle against AIDS, see M. D. Greenberg, "AIDS, Experimental Drug Approval, and the FDA New Drug Screening Process," *Legislation and Public Policy* 3, no. 295 (2000).

95   **Patients in merely "serious" trouble:** Frank E. Young, "Experimental Drugs for the Desperately Ill," *FDA Consumer* 21 (June 1987); Young, "The Role of the FDA in the Effort Against AIDS," 244–245.

95   **Washington now demonstrated its willingness:** Centers for Disease Control, "Notice to Readers: Pentamidine Methanesulfonate to Be Distributed by CDC," *Morbidity and Mortality Weekly Report* 33, no. 17 (May 4, 1984): 225–226; Centers for Disease Control, "Notice to Readers: U.S.-Manufactured Pentamidine Isethionate Cleared for Investigational Use," *Morbidity and Mortality Weekly Report* 33, no. 19 (May 18,1984): 270; "Wider Use of Pneumonia Drug Approved—Aerosolized Pentamidine," *FDA Consumer* 23, no. 3 (April 1989).

96   **In the late 1980s NIAID began funding:** Frank Young, "AIDS Research Comes to Patients' Home Towns," *FDA Consumer* 23, no. 4 (May 1989); Young, "Experimental Drugs for the Desperately Ill."

96   **By 1995, the FDA had granted treat-and-learn licenses:** Flieger, "FDA Finds New Ways to Speed Treatment to Patients."

97   **In 1964, Jacob Sheskin, an Israeli physician:** Rock Brynner and Trent Stephens, *Dark Remedy: The Impact of Thalidomide and Its Revival as a Vital Medicine* (New York: Basic Books, 2001), 122–123 (citing interviews conducted in 2000 with Dr. Gilla Kaplan and Dr. Robert Hastings). Brynner and Stephens describe the patient as a man. An earlier source, however, expressly notes that the patient was female: *Progress in Medicinal Chemistry* 22 (1985): 192.

97   **The PHS then extended that policy:** Brett Lowell, "Growing Interest in Thalidomide," *Treatment Issues* 9, no. 5 (May 1995); John S. James, "Thalidomide for Wasting Syndrome: Progress Toward Compromise," *AIDS Treatment News*, November 3, 1995.

97   **In 1995, the FDA, in an action:** Sheryl Gay Stolberg, "37 Years Later, a Second Chance for Thalidomide," *New York Times*, September 23, 1997.

98   **Before the license was issued, thalidomide's annual sales:** "History of Thalidomide," News-Medical.net, www.news-medical.net/health/History-of-Thalidomide.aspx (accessed January 16, 2013).

100   **"Official clinical research is usually years behind":** James, "Thalidomide for Wasting Syndrome."

CHAPTER 9 THE FADING MYTH
OF THE FDA'S "GOLD STANDARD"

103   **Statisticians call it the "reference class problem":** Jim Manzi, *Uncontrolled: The Surprising Payoff of Trial-and-Error for Business, Politics, and Society* (New York: Basic Books, 2012), 84.

105   **A recent report from the National Research Council:** Committee on a Framework for Developing a New Taxonomy of Disease, National Research Council, *Toward Precision Medicine: Building a Knowledge Network for Biomedical Research and a New Taxonomy of Disease* (Washington, DC: National Academies Press, 2011), 16.

105   **In 2003 and 2004, the FDA granted accelerated approval to two drugs:** John R. Johnson et al., "Approval Summary for Erlotinib for Treatment of Patients with Locally Advanced or Metastatic Non–Small Cell Lung Cancer after Failure of at Least One Prior Chemotherapy Regimen," *Clinical Cancer Research* 11, no. 18 (September 15, 2005): 6414–6421.

105   **After further trials failed to establish that Iressa extends:** Ed Silverman, "AstraZeneca Finally Gives Up on Iressa Approval," Pharmalot, February 9, 2011, www.pharmalot.com/2011/02/astrazeneca-finally-gives-up-on-iressa -approval (accessed January 24, 2013).

106   **We do already know that Iressa survival times:** Sarah Turner and Val Brickates Kennedy, "AstraZeneca's Iressa Gets E.U. Marketing Authorization," *Wall Street Journal*, July 1, 2009.

106   **One such patient who started on Iressa in 2004:** Ed Levitt, "Survivor's Story: Ed Levitt Surviving Against the Odds," *Network Connection*, Fall 2012, www.gacancersurvivors.org/sites/default/files/SurvivorStory_EdLevitt .pdf (accessed December 18, 2012); dog walking report based on personal communication with Paul Howard, Manhattan Institute, who spoke with the patient.

106   **As Dr. Janet Woodcock, the director:** Janet Woodcock, "A Framework for Biomarker and Surrogate Endpoint Use in Drug Development," PowerPoint presentation, November 4, 2004, slides 8, 36, www.fda.gov/ohrms/ dockets/ac/04/slides/2004-4079S2_03_Woodcock.ppt (accessed January 17, 2013).

107   **According to a recent consensus report issued:** Samir N. Khleif et al., "AACR-FDA-NCI Cancer Biomarkers Collaborative Consensus Report: Advancing the Use of Biomarkers in Cancer Drug Development," *Clinical Cancer Research* 16, no. 13 (July 1, 2010): 3299–3318.

108   **The agency points out that linking:** Russell Katz, "Biomarkers and Surrogate Markers: An FDA Perspective," *NeuroRx.* 1, no. 2 (April 2004): 189–195.

108   **So, as we have seen, the FDA won't approve:** Jeffrey Shuren, "Direct-to-Consumer Genetic Testing and the Consequences to the Public," statement before the Subcommittee on Oversight and Investigations, Committee on Energy and Commerce, and U.S. House of Representatives, July 22, 2010,

www.fda.gov/NewsEvents/Testimony/ucm219925.htm (accessed December 19, 2012).

108 **Over the past fifteen years, however:** Nicolaos Christodoulides et al., "Tools for Affordable Health Care," *Texas CEO Magazine*, July 31, 2011.

109 **In a tacit admission of the limits of its own trial protocols:** "International Serious Adverse Events Consortium (iSAEC)," FDA, November 1, 2010, www.fda.gov/AboutFDA/PartnershipsCollaborations/PublicPrivatePartnershipProgram/ucm231133.htm (accessed January 13, 2013).

109 **In 2010, the group released data:** U.S. Food and Drug Administration, "Critical Path Initiative: Report on Projects Receiving Critical Path Support," Fiscal Year 2010 Report (2010), 3, fda.gov/downloads/ScienceResearch/SpecialTopics/CriticalPathInitiative/UCM249262.pdf (accessed January 31, 2013).

110 **This gets us back to the 2011 NRC report:** National Research Council, *Toward Precision Medicine*.

111 **"What was interesting about Google Maps":** Quoted in Amy Dockser Marcus, "Working Toward a New Social Contract for Health Care," *Wall Street Journal*, April 17, 2012.

113 **In a 2011 essay:** Bruce A. Chabner, "Early Accelerated Approval for Highly Targeted Cancer Drugs," *New England Journal of Medicine* 364, no. 12 (March 24, 2011): 1087–1089.

113 **The recent drug-innovation report issued:** President's Council of Advisors on Science and Technology, "Report to the President on Propelling Innovation in Discovery, Development, and Evaluation," www.whitehouse.gov/sites/default/files/microsites/ostp/pcast-fda-final.pdf (accessed April 12, 2013).

CHAPTER 10 ADAPTIVE TRIALS

115 **As discussed in Chapter 9, a recent NRC report concludes:** Committee on a Framework for Developing a New Taxonomy of Disease, National Research Council, *Toward Precision Medicine: Building a Knowledge Network for Biomedical Research and a New Taxonomy of Disease* (Washington, DC: National Academies Press, 2011), 48.

116 **In the words of Dr. Raymond Woosley:** Quoted in Malorye Allison, "Reinventing Clinical Trials," *Nature Biotech* 30, no. 1 (January 9, 2012): 41–49.

116 **Nine months after the NRC issued its report:** President's Council of Advisors on Science and Technology, "Report to the President on Propelling Innovation in Discovery, Development, and Evaluation," www.whitehouse.gov/sites/default/files/microsites/ostp/pcast-fda-final.pdf (accessed April 12, 2013).

119 **Adaptive trials can be structured:** Allison, "Reinventing Clinical Trials."; Derek Lowe, "What You Need to Know About Adaptive Trials," *Pharmaceutical Executive*, July 1, 2006.

119   **The selection of additional biomarkers:** Anne R. Pariser, Kui Xu, John Milto, and Timothy R. Cote, "Regulatory Considerations for Developing Drugs for Rare Diseases: Orphan Designations and Early Phase Clinical Trials," *Discovery Medicine*, April 21, 2011; Thomas M. Burton, "Many 'Orphan Drugs' Get Expedited Review by FDA," *Wall Street Journal*, October 12, 2011. See also Edward R. Winstead, "Pancreatic Cancer Report Urges Changes in Clinical Trials," *NCI Cancer Bulletin* 6, no. 21 (November 3, 2009).

119   **Moreover, the analytical engines that quantify:** For one proposal along those lines, see Anup Malani, Oliver Bembom, and Mark van der Laan, "Accounting for Differences Among Patients in the FDA Approval Process," John M. Olin Law and Economics Working Paper no. 488, Public Law and Legal Theory Working Paper no. 281, University of Chicago School of Law, October 2009: 3, 8, www.law.uchicago.edu/files/file/488–281-am-fda.pdf (accessed January 17, 2013).

122   **NRC report cochair Dr. Susan Desmond-Hellmann:** Quoted in Amy Dockser Marcus, "Working Toward a New Social Contract for Health Care," *Wall Street Journal*, April 17, 2012.

122   **The ODA also gives the FDA the flexibility:** Pariser et al., "Regulatory Considerations for Developing Drugs for Rare Diseases"; Burton, "Many 'Orphan Drugs' Get Expedited Review by FDA."

123   **The FDA has designated as orphans:** President's Council of Advisors on Science and Technology, "Report to the President on Propelling Innovation in Discovery, Development, and Evaluation," iv.

123   **The orphanage currently fosters:** Walter Armstrong, "Pharma's Orphans," *Pharmaceutical Executive*, May 1, 2010.

123   **Experience with HIV and AIDS drugs:** Lowe, "What You Need to Know About Adaptive Trials."

CHAPTER 11 REASSEMBLING THE PIECES: PART 2

125   **In 1948, a century after John Snow:** Sharon Bertsch McGrayne, *The Theory That Would Not Die: How Bayes' Rule Cracked the Enigma Code, Hunted Down Russian Submarines, and Emerged Triumphant from Two Centuries of Controversy* (New Haven, CT: Yale University Press, 2011), 115 (and unless otherwise noted, the quotations in the following passages are from her book); William B. Kannel et al., "An Investigation of Coronary Heart Disease in Families," *American Journal of Epidemiology* 110, no. 3 (January 31, 1979): 281–290.

126   **"I saw a stegosaurus" is never believable:** Daniel M. Kammen, Alexander I. Shlyakhter, and Richard Wilson, "What Is the Risk of the Impossible?" Center for Domestic and Comparative Policy Studies Report, Woodrow Wilson School, Princeton University, 1993, 93–96, reprinted in *Journal of the Franklin Institute* (1995), http://users.physics.harvard.edu/~wilson/publications/ppaper544.html (accessed January 17, 2013).

126   **Indeed, the rise of modern Bayesian analysis began:** McGrayne, *The Theory That Would Not Die*, 130.

126   **As one Bayesian analyst put it: "The limit of [frequentist] approaches":**
       Quoted in McGrayne, *The Theory That Would Not Die*, 209.

127   **If you have no idea where that "of course" came from:** In a survey framed
       with somewhat different numbers so that a mere 9 out of 10 diagnoses are
       wrong, most American doctors estimated that 3 out of 4 would be right.
       Raymond Zhong, "Unreasonable Doubt," *Wall Street Journal*, October 14,
       2011. See Gerd Gigerenzer, *Adaptive Thinking: Rationality in the Real World*
       (New York: Oxford University Press, 2000), 65–66.

127   **Cornfield had helped design:** McGrayne, *The Theory That Would Not Die*,
       115–116.

129   **"Evaluation of drug safety has much in common":** Jerry Avorn, "Evalu-
       ating Drug Effects in the Post-Vioxx World: There Must Be a Better Way,"
       Circulation 113 (2006): 2173–2176.

129   **The FDA's "controlled" trials thus deliberately exclude:** For a statistician's
       view of the arcane details, see Anup Malani, Oliver Bembom, and Mark van
       der Laan, "Accounting for Differences Among Patients in the FDA Approval
       Process," John M. Olin Law and Economics Working Paper No. 488, Pub-
       lic Law and Legal Theory Working Paper No. 281, University of Chicago
       School of Law, October 2009, 3, 8, www.law.uchicago.edu/files/file/488–281
       -am-fda.pdf (accessed January 17, 2013).

130   **The EuResist analytical engine also takes into account:** Michal Rosen-Zvi et
       al., "Selecting Anti-HIV Therapies Based on a Variety of Genomic and Clin-
       ical Factors," *Bioinformatics* 24, no. 13 (2008): i399-i406; EuResist, "EuResist
       Prediction System," http://engine.euresist.org (accessed January 17, 2013).

130   **As the FDA's own Dr. Janet Woodcock:** Janet Woodcock, "A Framework
       for Biomarker and Surrogate Endpoint Use in Drug Development," Novem-
       ber 4, 2004, www.fda.gov/ohrms/dockets/ac/04/slides/2004-4079S2_03_
       Woodcock.ppt (accessed January 17, 2013).

131   **Today's computers, however, routinely perform half a million:** McGrayne,
       The Theory That Would Not Die, 219–220, 223–225.

131   **Bill Gates has attributed much of Microsoft's success:** Ibid., 242.

131   **Andy Grove, the pioneering founder:** Andy Grove, "Rethinking Clinical
       Trials," *Science* 333, no. 6050 (September 23, 2011): 1679.

132   **Their interests, *Wired* reported, include:** Daniela Hernandez, "Social
       Codes: Sharing Your Genes Online," *Wired: Science*, November 9, 2012.

132   **As we have seen, the FDA is determined to protect:** Letter from Alberto
       Gutierrez, Office of In Vitro Diagnostic Evaluation and Safety for the Cen-
       ter for Device and Radiological Health, to Anne Wojcicki, 23andMe Per-
       sonal Genome Service, "RE: 23andMe Personal Genome Service," June 10,
       2010, www.fda.gov/downloads/MedicalDevices/ResourcesforYou/Industry
       /UCM215240.pdf (accessed January 23, 2013).

133   **In February 2010, it did finally issue a draft guidance:** U.S. Food and Drug
       Administration, "Adaptive Design Clinical Trials for Drugs and Biologics,"
       FDA Guidance for Industry, February, 2010, www.fda.gov/downloads/Drugs
       / . . . /Guidances/ucm201790.pdf (accessed January 17, 2013).

133   **But it has clearly failed to proceed:** See, for example, "Chance for Change?" *Nature Reviews Drug Discovery* 5 (January 2006): 3.

133   **The PCAST report notes "widespread concern":** President's Council of Advisors on Science and Technology, "Report to the President on Propelling Innovation in Discovery, Development, and Evaluation," 51.

133   **One of the FDA's legitimate technical concerns is:** McGrayne, *The Theory That Would Not Die*, 228–229.

134   **"Far better an approximate answer":** Ibid., 233.

CHAPTER 12 ANTHRAXING WALL STREET

139   **It has ended up locking many manufacturers into the comparatively clumsy:** John E. Calfee and Scott Gottlieb, "Putting Markets to Work in Vaccine Manufacturing," *Health Outlook*, American Enterprise Institute for Public Policy Research (November 1, 2004), www.aei.org/article/health/putting-markets-to-work-in-vaccine-manufacturing (accessed January 9, 2013); Lawrence Corey, "Thirty Years of Fighting Aids: A Progress Report," *Wall Street Journal*, June 18, 2011.

139   **Another study, published in 1998 in the same journal:** Fiona Godlee, "Wakefield's Article Linking MMR Vaccine and Autism Was Fraudulent," *British Medical Journal* 342 (March 15, 2011): 64–66; Paul A. Offit, *Autism's False Prophets: Bad Science, Risky Medicine, and the Search for a Cure* (New York: Columbia University Press, 2010).

139   **Twelve years after it helped launch:** Editors of the *Lancet*, "Retraction— Ileal-Lymphoid-Nodular Hyperplasia, Non-Specific Colitis, and Pervasive Developmental Disorder in Children," *Lancet* 375, no. 9713 (February 6, 2010): 445; Nathan Seppa, "Journal Retracts Flawed Study Linking MMR Vaccine and Autism," *Science News*, February 3, 2010.

139   **In 1985—just as the tools of intelligent design:** Division of Health and Promotion and Disease Prevention, National Research Council, *Vaccine Supply and Innovation* (Washington, DC: National Academies Press, 1985).

140   **Plaintiffs were also suing Washington:** *Berkovitz v. United States*, 486 US 531 (1988).

140   **In early 2011, in a case involving a vaccine:** *Bruesewitz v. Wyeth LLC*, 562 U.S. __, 131 S.Ct.1068 (2011).

140   **Today Washington directly or indirectly pays:** "Vaccines—Endangered Species?" *Nature Immunology* 3, no. 695 (2002).

140   **As Washington took control of almost all the science:** Calfee and Gottlieb, "Putting Markets to Work in Vaccine Manufacturing."

140   **The price of the vaccine, if approved, would be determined:** Lance E. Rodewald et al., "Vaccine Supply Problems: A Perspective of the Centers for Disease Control and Prevention," *Clinical Infectious Diseases* 42, no. 3 (2006): S104–S110. See also Awi Federgruen, "The Drug Shortage Debacle—and How to Fix It," *Wall Street Journal*, March 1, 2012.

141   **More than a dozen companies were manufacturing:** Calfee and Gottlieb, "Putting Markets to Work in Vaccine Manufacturing"; Division of Health

and Promotion and Disease Prevention, National Research Council, *Vaccine Supply and Innovation*.

141   **In 2004, Dr. Scott Gottlieb, soon to be appointed:** Calfee and Gottlieb, "Putting Markets to Work in Vaccine Manufacturing."

141   **A 2002 letter to a leading medical journal:** David M. Shlaes and Robert C. Moellering Jr., "The United States Food and Drug Administration and the End of Antibiotics," *Clinical Infectious Diseases* 34, no. 3 (February 1, 2002): 420–422.

141   **"Without significant changes from the FDA":** David M. Shlaes, *Antibiotics: The Perfect Storm* (New York: Springer, 2010): 28.

142   **While Pfizer was still charging Americans $10 to $30:** Donald G. McNeil, Jr., "Medicine Merchants: Patents and Patients; As Devastating Epidemics Increase, Nations Take On Drug Companies," *New York Times*, July 9, 2000.

142   **In Norway, only a tiny number:** Steve Jones, *Darwin's Ghost: The Origin of Species Updated* (New York: Ballantine, 2000), 89.

142   **As Steve Jones notes:** Ibid.

142   **In the balance between abstract concerns:** See, for example, Josh Bloom, "Where Is Mel Blanc When You Need Him?" *Medical Progress Today*, November 9, 2011.

143   **Then, for a long stretch that began soon after:** Abigail Colson, "The Antibiotic Pipeline," Extending the Cure: Policy Responses to the Growing Threat of Antibiotic Resistance, Policy Brief ETC-06, May 2008, www.rff .org/RFF/Documents/ETC-06.pdf (accessed January 12, 2013).

143   **Encouraged by the resurgent germs, the antibiotic industry:** Brian Vastag, "NIH Superbug Outbreak Highlights Lack of New Antibiotics," *Washington Post*, August 24, 2012.

144   **A gonorrhea vaccine "remains key":** Quoted in Jason Koebler, "CDC Warns Untreatable Gonorrhea Is on the Way," *U.S. News*, February 13, 2012.

145   **After the 2001 attacks, however, three manufacturers immediately offered:** Glenn Hess, "US Ponders Bayer 'Cipro' Patent, Generic Anthrax Options," *ICIS News*, October 17, 2001.

145   **Congress leaped into action and enacted a sweeping:** The Project BioShield Act of 2004, 42 U.S.C. 243 et seq.

145   **No human trials are required, and a "lower level of evidence":** "Emergency Use Authorization of Medical Products," U.S. Food and Drug Administration, July 2007, www.fda.gov/RegulatoryInformation/Guidances /ucm125127 (accessed January 12, 2013).

145   **As the ink was drying on the new law, a federal judge agreed:** *Doe v. Rumsfeld*, 297 F. Supp. 2d 119 (D.D.C. 2003); *Doe v. Rumsfeld*, 341 F. Supp. 2d 1 (D.D.C. 2004).

145   **So the Pentagon invoked the Cessna law:** See U.S. Department of Homeland Security and United States Coast Guard, "Commandant Instruction M6230.3B," September 10, 2007, www.uscg.mil/directives/cim/6000–6999 /CIM_6230_3B.pdf (accessed January 12, 2013); "Judge Sullivan's April 6, 2005 Ruling on *John Doe et al vs. Donald Rumsfeld et al.*," Military Biodefense Vaccine Project, April 6, 2005, www.military-biodefensevaccines.org

/documents/sullivan.htm (accessed January 12, 2013); Stuart L. Nightingale et al., "Emergency Use Authorization (EUA) to Enable Use of Needed Products in Civilian and Military Emergencies, United States," *Emerging Infectious Disease* 13, no. 7 (July 2007): 1046–1051.

145   **At the same time, ever so quietly, blanket liability protection:** "HHS Announces New Steps in Anthrax Preparedness," U.S. Department of Health and Human Services news release, October 1, 2008, www.hhs.gov/news /press/2008pres/10/20081001a.html (accessed January 12, 2013).

145   **In the last decade Washington has funded the development:** "Scientists Report Progress Making New Anthrax Vaccine," *Lodi News-Sentinel*, September 28, 2006; Associated Press, "New Anthrax Vaccine Shows Promise," September 27, 2006; Robert Roos, "VaxGen Sells Anthrax Vaccine to Rival Firm," Center for Infectious Disease Research and Policy News, May 7, 2008, www.cidrap.umn.edu/cidrap/content/bt/anthrax/news/may0708 anthrax.html (accessed May 15, 2013).

CHAPTER 13 DOLLAR DOCTOR SCIENCE

147   **The CEO of Novartis switched to Pfizer's:** Jeanne Whalen, "Novartis CEO Vasella: Satisfied Lipitor Customer," *Wall Street Journal*, January 3, 2008.

148   **The former director of President Obama's:** Quoted in Jerome Groopman, "Health Care: Who Knows 'Best'?" *New York Review of Books*, February 11, 2010; Peter R. Orszag, "Statement on Increasing the Value of Federal Spending on Health Care," testimony to Committee on the Budget, U.S. House of Representatives, July 16, 2008, www.cbo.gov/sites/default/files/cbofiles/ftp docs/95xx/doc9563/07–16-healthreform.pdf (accessed April 15, 2013).

148   **A single committee, the U.S. Preventive Services Task Force:** See Scott Gottlieb, "Meet the ObamaCare Mandate Committee," *Wall Street Journal*, February 16, 2012.

149   **Dr. Griswold must surrender control:** Jeffery A. Singer, "The Coming Medical Ethics Crisis," Reason.com, March 15, 2012, http://reason.com/archives /2012/03/15/the-coming-medical-ethics-crisis (accessed January 12, 2013).

149   **A federal judge struck down these conditions as unconstitutional:** *Washington Legal Foundation v. Friedman*, 13 F. Supp.2d 51 (D.D.C. 1998).

150   **In late 2012 another appellate court upheld:** *United States v. Caronia*, 703 F.3d 149 (2d Cir. 2012).

150   **He pled guilty to a misdemeanor:** Harvey Silverglate, "A Doctor's Posthumous Vindication," *Wall Street Journal*, December 25, 2012.

152   **In an article published in early 2010, Dr. Jerome Groopman:** Jerome Groopman, "Health Care: Who Knows 'Best'?" *New York Review of Books*, February 11, 2010. See also his *How Doctors Think* (New York: Mariner Books, 2008).

152   **"In the case of prostate cancer":** Leonard A. Zwelling, "'Comparative Effectiveness' Research Is Always Behind the Curve," *Wall Street Journal*, March 16, 2010.

153   **Washington's Preventive Services committee usually demands:** Gottlieb, "Meet the ObamaCare Mandate Committee."

153   **In early 2009 a federal judge appointed:** *Tummino v. Torti*, 603 F.Supp. 2d 519 (E.D.N.Y. 2009).

154   **"Supporters and opponents both said":** Rob Stein, "Controversial 'Ella' Contraceptive Now Available in U.S. for the First Time," *Washington Post*, December 1, 2010.

154   **The Obama White House quickly denied any involvement:** Jennifer Corbett Dooren, "Obama Health Chief Blocks FDA on 'Morning After' Pill," *Wall Street Journal*, December 8, 2011.

154   **In April 2013, describing the administration's action:** *Tummino v. Hamburg* __ F.Supp. 2d __, 2013 WL 1348656 (E.D.N.Y. April 5, 2013).

154   **In the interim, Washington had notified Belmont Abbey:** See Charlotte Allen, "The Persecution of Belmont Abbey," *Weekly Standard*, October 26, 2009.

154   **The average first-birth maternal age in the United States:** T. J. Mathews and Brady E. Hamilton, "Delayed Childbearing: More Women Are Having Their First Child Later in Life," *NCHS Data Brief*, no. 21 (August 2009).

154   **A 2009 statement by Britain's Royal College:** "RCOG Statement on Later Maternal Age," Royal College of Obstetricians and Gynaecologists, June 15, 2009, www.rcog.org.uk/what-we-do/campaigning-and-opinions /statement/rcog-statement-later-maternal-age (accessed January 12, 2013).

155   **"to provide socially sustainable, cost-effective care":** Ezekiel J. Emanuel and Victor R. Fuchs, "The Perfect Storm of Overutilization," *Journal of the American Medical Association* 299, no. 23 (2008): 2789–2791; Govind Persad, Alan Wertheimer, and Ezekiel J. Emanuel, "Principles for Allocation of Scarce Medical Interventions," *Lancet* 373, no. 9661 (January 31, 2009): 426–431.

155   **In March 2011, when the FDA approved:** Sten Stovall, "Lupus Drug Suffers U.K. Setback," *Wall Street Journal*, September 30, 2011.

157   **That will take us to the threshold of what many critics:** Marcia Angell, *The Truth About the Drug Companies: How They Deceive Us and What to Do About It* (New York: Random House Trade Paperbacks, 2005), 253, 255.

CHAPTER 14 THE RISING COST OF HELPLESS CARE

159   **The "major contributor" to rapidly rising health care costs:** C. C. Denny et al., "Why Well-Insured Patients Should Demand Value-Based Insurance Benefits," *Journal of the American Medical Association* 297, no. 22 (2007): 2515–2518.

161   **But as Nobel economist Robert Fogel:** Robert Fogel, "Forecasting the Cost of U.S. Healthcare," *The American*, September 3, 2009.

163   **Biochemists are now able to construct inert:** Avi Schroeder et al., "Remotely Activated Protein-Producing Nanoparticles," *Nano Letters* 12, no. 6 (March 20, 2012): 2685–2689.

163   **"If we know which genes control longevity":** Quoted in Ruth Elaine Nieuwenhuis-Mark, "Healthy Aging as Disease?" *Frontiers in Aging Neuroscience* 3, no. 3 (February 22, 2011): 1.

163   **As Fogel points out, most of the growth:** Robert Fogel, "Forecasting the Cost of U.S. Healthcare."

163   **The late British economist Angus Maddison estimated:** Michael Milken, "Health-Care Investment—the Hidden Crisis," *Wall Street Journal*, February 8, 2011.

163   **According to one 2006 estimate, Americans valued:** Kevin M. Murphy and Robert H. Topel, "The Value of Health and Longevity," *Journal of Political Economy* 114, no. 5 (October 2006): 871–904.

165   **Typical of countless others in the field, one 2006 study:** Karen E. Lasser et al., "Access to Care, Health Status, and Health Disparities in the United States and Canada: Results of a Cross-National Population-Based Survey," *American Journal of Public Health: Research and Practice* 96, no. 7 (July 2006): 1300–1307.

165   **"We pay almost twice what Canada does":** Quoted in Maggie Fox, "Canadians Healthier than Americans—Study," Reuters, May 31, 2006.

165   **"This finding," according to a second author:** Quoted in Physicians for National Health Care, "U.S. Residents Less Healthy than Canadians," People's World, June 9, 2006, http://peoplesworld.org/u-s-residents-less-healthy-than-canadians (accessed January 14, 2013).

165   **"This speaks more to genetics, diet, exercise and culture":** David Gratzer, "Where Would You Rather Be Sick?" *Wall Street Journal*, June 15, 2006.

165   **All things considered, there's "not an iota of evidence":** David U. Himmelstein and Steffie Woolhandler, "Healthy Debates Engage U.S. and Canada," *Wall Street Journal*, June 22, 2006.

165   **We do know that U.S. obesity rates:** Katherine M. Flegal et al., "Prevalence of Obesity and Trends in the Distribution of Body Mass Index Among US Adults, 1999–2010," *Journal of the American Medical Association* 307, no. 5 (February 1, 2012): 491–497.

165   **But according to Washington's own statisticians, "the primary reason":** Marian F. MacDorman and T. J. Mathews, "Behind International Rankings of Infant Mortality: How the United States Compares with Europe," *NCHS Data Brief* 23 (November 2009).

166   **Most neonatal deaths occur very soon after birth:** Scott W. Atlas, "Survival of the Smallest," *Hoover Digest*, January 23, 2012; Scott W. Atlas, *In Excellent Health: Setting the Record Straight on America's Health Care* (Stanford, CA: Hoover Institution Press, 2012).

166   **We have these facts on the authority of "Eight Americas":** Christopher J. L. Murray et al., "Eight Americas: Investigating Mortality Disparities Across Races, Counties, and Race-Counties in the United States," *PLoS Medicine* 3, no. 9 (September 2006): e260.

167   **One of the study's authors ventured to suggest that where you live:** Associated Press, "Where You Live Can Affect How Long You Live," NPR, Sep-

tember 11, 2006, www.npr.org/templates/story/story.php?storyId=6057076 (accessed January 23, 2013).

CHAPTER 15
THE FALLING COST OF HEALTH CARE

171   **In his April 2005 obituary, the *New York Times*:** Lawrence K. Altman, "Obituary: Maurice Hilleman, Vaccine Creator," *New York Times*, April 13, 2005.

172   **Even as the population has aged, hospital admissions:** Ron Winslow and Shirley S. Wang, "Heart Failure Puts Fewer in Hospital," *Wall Street Journal*, October 19, 2011.

172   **Nurses and pharmacists already provide:** R. Pete Vanderveen, "How to Care for 30 Million More Patients," *Wall Street Journal*, July 19, 2010.

173   **But when penicillin manufactured by Merck:** John S. Mailer Jr. and Barbara Mason, "Penicillin: Medicine's Wartime Wonder Drug and Its Production at Peoria, Illinois," Illinois Periodicals Online, www.lib.niu.edu/2001/iht810139.html (accessed January 13, 2013).

173   **Kary Mullis:** Celia Farber, "Interview with Kary Mullis," *Spin Magazine*, July 1994, www.reviewingaids.com/awiki/index.php/Document:Farber_interviews_Mullis (accessed January 13, 2013).

176   **According to one estimate, getting a drug to market:** Jonah Lehrer, "Trials and Errors: Why Science Is Failing Us," *Wired*, December 16, 2011.

177   **In 2005, Genentech announced the successful results:** "Genentech Restricts Use of Avastin for Ophthalmic Use," American Macular Degeneration Foundation, www.macular.org/noavas.html (accessed January 13, 2013); Andrew Pollack, "Genentech Will Restrict Eye Use of Cancer Drug," *New York Times*, October 11, 2007.

177   **As of 2011, Washington was poised to begin:** National Eye Institute, "NIH Study Finds Avastin and Lucentis Are Equally Effective in Treating Age-Related Macular Degeneration," NIH News, April 28, 2011, www.nih.gov/news/health/apr2011/nei-28.htm (accessed January 13, 2013).

177   **Hospitals and doctors have been reporting shortages:** Anna Yukhananov, "Authorities Perplexed by Drug Shortage Spike," Reuters, October 14, 2011.

178   **At the end of 2011, a record 267 prescription drugs:** Linda A. Johnson, "2011 Medication Shortages Set New Record at 267," *Journal Record*, January 3, 2012; Scott Gottlieb, "Solving the Growing Drug Shortages," *Wall Street Journal*, November 4, 2011.

178   **She also identifies the problem:** Marcia Angell, *The Truth About the Drug Companies: How They Deceive Us and What to Do About It* (New York: Random House Trade Paperback, 2005), 92.

179   **A 2006 article in the *New England Journal of Medicine*:** Alastair J. J. Wood, "A Proposal for Radical Changes in the Drug-Approval Process," *New England Journal of Medicine* 355 (August 10, 2006): 618–623.

CHAPTER 16 REASSEMBLING THE PIECES: PART 3

181   **The original patents for eflornithine:** Robert S. Desowitz, *Federal Bodysnatchers and the New Guinea Virus: Tales of Parasites, People, and Politics* (New York: W. W. Norton, 2004), 142–144.

182   **As told by the *New York Times* journalist Tina Rosenberg:** Tina Rosenberg, "The Scandal of 'Poor People's Diseases,'" *New York Times*, March 29, 2006, http://new.tballiance.org/newscenter/view-innews.php?id=605 (accessed January 16, 2013).

185   **As epidemiologist Robert Desowitz recounts:** Desowitz, *Federal Bodysnatchers and the New Guinea Virus*, 143–144.

186   **The patent law covers "methods" as well:** *Mayo Collaborative Services v. Prometheus Laboratories, Inc.*, 566 U.S. ___, 132 S.Ct. 1289 (2012); *Ass'n for Molecular Pathology v. Myriad Genetics* __U.S. __ (No. 12-398, June 13, 2013). See also *Ariad Pharmaceuticals, Inc. v. Eli Lilly & Co.*, 598 F.3d 1336, 1341 (Fed. Cir. 2010).

186   **Myriad Genetics' patent claims associated with two breast cancer genes:** *Ass'n for Molecular Pathology v. Myriad Genetics* __U.S. __ (No. 12-398, June 13, 2013).

186   **As the courts have recognized:** *Mayo Collaborative Services v. Prometheus Laboratories, Inc.*, 566 U.S. __, 132 S.Ct. 1289 (2012); *Ass'n for Molecular Pathology v. Myriad Genetics*.

187   **The copycat manufacturer can't even be sued:** *PLIVA v. Mensing*, 564 U.S. ___, 131 S. Ct. 2567 (2011); see also *Mutual Pharmaceutical Co. v. Bartlett*, No. 12-142 (U.S. June 24, 2013), 570 U.S. ___ (2013).

188   **Some courts have decided that the pioneer may be sued:** *Conte v. Wyeth, Inc.*, 85 Cal. Rptr. 3d 299 (Ct. App. 2008); *Kellogg v. Wyeth*, 762 F. Supp. 2d 694. (D. Vt. 2010).

188   **The 2011 NRC report:** Committee on a Framework for Developing a New Taxonomy of Disease, National Research Council, *Toward Precision Medicine: Building a Knowledge Network for Biomedical Research and a New Taxonomy of Disease* (Washington, DC: National Academies Press, 2011), 58.

188   **The co-chair of the task force has independently suggested:** Amy Dockser Marcus, "Working Toward a New Social Contract for Health Care," *Wall Street Journal*, April 17, 2012. See also Susan Desmond-Hellmann, "Toward Precision Medicine: A New Social Contract?" *Science Translational Medicine* 4, no. 129 (April 11, 2012): 129ed3.

189   **But affirming the private ownership:** See also Alastair J. J. Wood, "A Proposal for Radical Changes in the Drug-Approval Process," *New England Journal of Medicine* 355 (August 10, 2006): 618–623. Wood outlines a proposal to extend data exclusivity rights to promote the ongoing study of safety and efficacy after a drug is licensed, the development of solid links between biomarkers and high-level clinical symptoms, and the development of drugs to treat a short list of common, currently untreatable diseases, selected by an independent committee—Alzheimer's or osteoarthritis, for example.

190 **And as we have seen, President Obama's own expert advisory council:** President's Council of Advisors on Science and Technology, "Report to the President on Propelling Innovation in Discovery, Development, and Evaluation," 48, www.whitehouse.gov/sites/default/files/microsites/ostp/pcast -fda-final.pdf (accessed on April 12, 2013).

190 **In 2005, a Rand Corporation analysis:** James H. Bigelow, Kateryna Fonkych, Constance Fung, and Jason Wang, *Analysis of Healthcare Interventions That Change Patient Trajectories* (Santa Monica, CA: Rand Corporation, 2005); Reed Abelson and Julie Creswell, "In Second Look, Few Savings from Digital Health Records," *New York Times*, January 10, 2013; Nick Triggle, "Pull Plug on NHS E-Records—MPs," BBC News Health, August 2, 2011, www.bbc.co.uk/news/health-14378346 (accessed January 17, 2013); "IT Firm Behind Unworkable NHS Database Keeps IT deal," Health Direct, February 22, 2012, www.healthdirect.co.uk/2012/02/it-firm-behind -unworkable-nhs-database-keeps-it-deal.html (accessed January 17, 2013).

192 **Rules drafted in the HIV-driven 1980s:** See Frank E. Young, "Experimental Drugs for the Desperately Ill," *FDA Consumer* 21 (June 1, 1987).

193 **In an isolated fishing village in Venezuela:** Erik Van Eaton, "Researcher's Experience Spurs Search for Fatal Gene," *Daily of the University of Washington*, October 20, 1995, http://dailyuw.com/archive/1995/10/20/imported /researchers-experience-spurs-search-fatal-gene#.UPgK967-uSp (accessed January 17, 2013).

193 **In another genetically isolated community in the Andes:** Pam Belluck, "Alzheimer's Stalks a Colombian Family," *New York Times*, June 1, 2010.

193 **"Two of the most daunting bottlenecks":** Desmond-Hellmann, "Toward Precision Medicine: A New Social Contract?"

CHAPTER 17 DYING ALONE

197 **"In the Victorian version of the Puritan ethic":** Gertrude Himmelfarb, *The Moral Imagination: From Edmund Burke to Lionel Trilling* (Chicago: Ivan R. Dee, 2006), 49.

198 **That problem, they agreed, could be handled in other ways:** *Griswold v. Connecticut*, 381 U.S. 479 (1965).

199 **To judge by the appalling trends:** "Incidence, Prevalence, and Cost of Sexually Transmitted Infections in the United States," CDC Fact Sheet (February 2013), www.cdc.gov/std/stats/STI-Estimates-Fact-Sheet-Feb-2013.pdf (accessed April 15, 2013).

200 **Pregnancy discrimination is illegal:** *Automobile Workers v. Johnson Controls, Inc.*, 499 U.S. 187 (1991).

200 **The mentally ill have a broad right:** *Olmstead v. L.C.*, 527 U.S. 581 (1999).

201 **Some gay-rights advocates worry instead:** Amy Harmon, "That Wild Streak? Maybe It Runs in the Family," *New York Times*, June 15, 2006.

201 **Prenatal testing now gives parents the power:** Lindsey Tanner, "Physicians Could Make the Perfect Imperfect Baby," *Los Angeles Times*, December 31, 2006.

202   **While First Lady Michelle Obama was denouncing:** Michael Moss, "While Warning About Fat, U.S. Pushes Cheese Sale," *New York Times*, November 6, 2010.

203   **After years of recommending diet and exercise:** Gina Kolata, "When Advice on Diabetes Is Sound, but Ignored," *New York Times*, October 17, 2006.

203   **"Whatever your thoughts are":** Quoted in ibid.

203   **In 2004, the FDA issued a report on "the widening gap":** U.S. Food and Drug Administration, "FDA's Critical Path Initiative," FDA.gov, December 28, 2012, www.fda.gov/ScienceResearch/SpecialTopics/CriticalPathInitiative /ucm076689.htm (accessed January 17, 2013).

203   **Some find their way to doctors who concoct:** Elizabeth Bernstein, "A New Breed of 'Diet' Pills," *Wall Street Journal*, August 22, 2006.

205   **After a long period in which vaccination rates fell:** Paul A. Offit, "Fatal Exemption," *Wall Street Journal*, January 20, 2007; Philip J. Smith et al., "Parental Delay or Refusal of Vaccine Doses, Childhood Vaccination Coverage at 24 Months of Age, and the Health Belief Model," *Public Health Rep* 126, suppl. 2 (2011): 135–146, www.ncbi.nlm.nih.gov/pmc/articles/PMC 3113438 (accessed January 17, 2013).

205   **In the words of Paul Offit:** Offit, "Fatal Exemption."

CHAPTER 18 THE RIGHT TO SNIFF

209   **One MIT researcher, for example, recently set up:** Robert Lee Hotz, "The Really Smart Phone," *Wall Street Journal*, April 22, 2011.

210   **But most manufacturers shy away from sharing:** Amy Dockser Marcus and Christopher Weaver, "Heart Gadgets Test Limits of Privacy-Law Limits," *Wall Street Journal*, November 28, 2012.

211   **These behavioral judgments can't be anchored:** Stephanie Pappas, "Scrutiny of Personal Gene Tests Increases," LiveScience, May 30, 2010, www .livescience.com/9927-scrutiny-personal-gene-tests-increases.html (accessed January 16, 2013).

211   **"There is no need for one":** Quoted in Sarah A. Leavitt, "'A Private Little Revolution': The Home Pregnancy Test in American Culture," *Bulletin of the History of Medicine* 80, no. 2 (Summer 2006): 326.

211   **The FDA didn't even begin seriously considering:** Warren E. Leary, "In Switch, U.S. Studies Home Test Kits for AIDS," *New York Times*, April 26, 1990.

212   **A close match with an innocent relative's DNA:** Turi E. King and Mark A Jobling, "What's in a Name? Y Chromosome, Surnames and the Genetic Genealogy Revolution," *Trends in Genetics* 25, no. 8 (August 2009): 351–360.

212   **Ethnic origin markers in DNA scans:** Gautam Naik, "To Sketch a Thief: Genes Draw Likeness of Suspects," *Wall Street Journal*, March 27, 2009; "DNA Left at Crime Scene Could Be Used to Create Picture of Criminal's Face," *Daily Mail*, February 17, 2009.

213 **And as we have seen, courts have recently been affirming:** *United States v. Caronia*, 703 F.3d 149 (2d Cir. 2012).

214 **A constitutional right to sniff can be affirmed without:** See, e.g., *Abigail Alliance for Better Access to Developmental Drugs v. von Eschenbach*, 495 F.3d 695 (D.C. Cir. 2007) (en banc), *cert. denied*, 128 S. Ct. 1069 (2008).

214 **Our grandchildren "will likely engineer themselves":** Juan Enriquez, "Homo Evolutis," Edge, www.edge.org/annual-question/2009/response/10179 (accessed January 16, 2013).

214 **Students use Ritalin and Provigil:** Henry Greely et al., "Toward Responsible Use of Cognitive-Enhancing Drugs by the Healthy," *Nature* 456 (December 11, 2008): 702–705.

215 **Redirected against fond memories:** Merel Kindt et al., "Beyond Extinction: Erasing Human Fear Responses and Preventing the Return of Fear," *Nature Neuroscience* 12 (February 15, 2009): 256–258; Catharine Paddock, "Beta-Blocker Erases Bad Memories," *Medical News Today*, February 16, 2009; Jonah Lehrer, "The Forgetting Pill Erases Painful Memories Forever," *Wired*, February 17, 2012.

CHAPTER 19 THE END OF SOCIALIZED MEDICINE

219 **By 2008, seven of the world's ten most profitable:** Pharmaceutical Executive Staff, "The Pharm Exec 50," *Pharmaceutical Executive*, May 2009: 68–78.

219 **The first patient we meet:** Sherwin B. Nuland, *How We Die: Reflections on Life's Final Chapters* (New York: Vintage Books, 1995), 3.

219 **In 1985, in their Nobel lecture on the lipid:** Michael S. Brown and Joseph L. Goldstein, "A Receptor-Mediated Pathway for Cholesterol Homeostasis," Nobel Prize Lecture, December 9, 1985, www.nobelprize.org /nobel_prizes/medicine/laureates/1985/brown-goldstein-lecture.pdf (accessed January 15, 2013).

220 **As Nuland notes, "The best assurance of longevity":** Nuland, *How We Die*, 76.

221 **Dr. David Kessler, the FDA's head at the time:** Quoted in Gina Kolata, "Vitamin to Protect Fetuses Will Be Required in Foods," *New York Times*, March 1, 1996.

221 **The public cure for a nutritional deficiency:** "Trends in Spina Bifida, United States, 1991–2005," Centers for Disease Control and Prevention, www.cdc.gov/Features/dsSpinaBifida (accessed January 15, 2013); "Trends in Folic Acid Supplement Intake Among Women of Reproductive Age-California, 2002–2006," *Morbidity and Mortality Weekly Report* 56, no. 42 (October 26, 2007): 1106–1109; Darshak M. Sanghavi, "A Growing Debate over Folic Acid in Flour," *New York Times*, December 4, 2007,.

223 **Britain, with one of the world's worst cholesterol problems:** See "Statins," Medical Research Council, October 2007, www.mrc.ac.uk/Achievements impact/Storiesofimpact/Statins/index.htm (accessed January 15, 2013).

225 **In the early 1920s Quaker Oats offered $900,000:** "Wisconsin Alumni Research Foundation History," Funding Universe, www.fundinguniverse .com/company-histories/Wisconsin-Alumni-Research-Foundation-company -History.html (accessed January 15, 2013).

225 *Consumer Reports* **estimated:** "New Generic Statins Mean Big Savings for Consumers Needing Cholesterol Reduction," *Consumers Union*, June 22, 2006, www.consumersunion.org/pub/core_health_care/003557.html (accessed January 15, 2013).

CHAPTER 20 THE CULTURE OF LIFE

227 **In Washington and elsewhere, authorities scrambled to block:** Denise Grady and William J. Broad, "Seeing Terror Risk, U.S. Asks Journals to Cut Flu Study Facts," *New York Times*, December 20, 2011.

227 **Germs engineered to track ethnic, racial:** Frank Herbert, *The White Plague* (New York: Tor, 2007).

227 **As Bill Joy, the founder of Sun:** Bill Joy, "Why the Future Doesn't Need Us," *Wired*, April 2000.

228 **A drug-resistant pneumonia that surfaced in Spain:** Steve Jones, *Darwin's Ghost: The Origin of Species Updated* (New York: Ballantine, 2000), 92.

228 **The hardy new form of the E. *coli* bacterium:** William Neuman and Scott Sayare, "Egyptian Seeds Are Linked to E. Coli in Germany and France," *New York Times*, June 30, 2011.

228 **Multicellular life's long flight from infectious germs:** Eleonora Market and F. Nina Papavasiliou, "V(D)J Recombination and the Evolution of the Adaptive Immune System," *PLoS Biology* 1, no. 1 (October 13, 2003): 24–27.

228 **Bacteria don't undergo "replicative aging":** Or maybe they do. Detailed genetic scrutiny has recently suggested that germs, too, undergo very slow "replicative aging." See Jose M. Gomez, "Aging in Bacteria, Immortality or Not—A Critical Review," *Current Aging Science* 3, no. 3 (December 2010): 198–218.

228 **That we have recently had serious trouble:** John E. Calfee and Scott Gottlieb, "Putting Markets to Work in Vaccine Manufacturing," *Health Policy Outlook*, American Enterprise Institute for Public Policy Research (November 1, 2004), www.aei.org/article/health/putting-markets-to-work-in-vaccine -manufacturing (accessed January 9, 2013).

230 **In the years during which all the key tools:** Henry G. Grabowski and Y. Richard Wang, "The Quantity and Quality of Worldwide New Drug Introductions, 1982–2003," *Health Affairs* 25, no. 2 (March 2006): 452–460.

230 **With digital and biochemical technologies:** Edmund S. Phelps, "Dynamic Capitalism," *Wall Street Journal*, October 10, 2006.

230 **In this regard, as the late Harvard economist:** Quoted in ibid.

231 **A 2011 study, commissioned by Pharma:** Battelle Technology Partnership Practice, "The U.S. Biopharmaceuticals Sector: Economic Contribu-

tion to the Nation," July 2011, www.phrma.org/sites/default/files/159/2011_battelle_report_on_economic_impact.pdf (accessed January 15, 2013).

234 **The culture of life compassionately respects:** Ken Murray, "Why Doctors Die Differently," *Wall Street Journal*, February 25, 2012.

234 **In June 2011 Britain launched a clinical trial:** Ben Hirschler, "Tests Start on HIV Biotech Drug—Grown in Tobacco," Reuters, July 19, 2011; Jeffrey L. Fox, "HIV Drugs Made in Tobacco," *Nature Biotechnology* 29, no. 10 (2011): 852.

234 **In April 2012 oncologists at the Children's Hospital:** "First Pediatric Patients Treated in T Cell Therapy Clinical Trial (CTL019, formerly CART19)," Children's Hospital of Philadelphia, www.chop.edu/service/oncology/pediatric-cancer-research/t-cell-therapy.html (accessed January 15, 2013); Catharine Paddock, "Child's Leukemia Cured by Her Own Re-Engineered Immune Cells," *Medical News Today*, December 12, 2012; Denise Gray, "In Girl's Last Hope, Altered Immune Cells Beat Leukemia," *New York Times*, December 10, 2012, A1; Claire Bates, "The Cancer Girl Cured by the HIV Virus: Seven-Year-Old Makes Extraordinary Recovery After U.S. Doctors Re-wire Her Immune System to Destroy Leukaemia Cells," *Daily Mail*, December 11, 2012.

INDEX